The Pay, Promotion, and Retention of High-Quality Civil Service Workers in the Department of Defense

Beth J. Asch

Prepared for the
Office of the Secretary of Defense

National Defense Research Institute

RAND

This study was sponsored by the Office of Civilian Personnel Policy of the Under Secretary of Defense for Personnel and Readiness. It was prepared within the Forces and Resources Policy Center of RAND's National Defense Research Institute (NDRI), a federally funded research and development center sponsored by the Office of the Secretary of Defense, the Joint Staff, the unified commands, and the defense agencies under contract DASW01-95-C-0059.

Library of Congress Cataloging-in-Publication Data

Asch, Beth J.
    The pay, promotion, and retention of high-quality civil service workers in the
    Department of Defense / Beth J. Asch.
        p. cm.
    MR-1193
    Includes bibliographical references.
    ISBN 0-8330-2988-6
    1. United States—Armed Forces—Civilian employees—Salaries, etc. 2. United
    States—Armed Forces—Civilian employees—Promotions. 3. United States. Dept.
    of Defense—Employees—Salaries, etc. 4. United States—Armed Forces—
    Personnel management. 5. United States. Dept. of Defense—Employees—
    Promotions. I. Title.

    UB193 .A83 2001
    355.6'19—dc21

                                                            2001019635

RAND is a nonprofit institution that helps improve policy and decisionmaking through research and analysis. RAND® is a registered trademark. RAND's publications do not necessarily reflect the opinions or policies of its research sponsors.

Published 2001 by RAND
1700 Main Street, P.O. Box 2138, Santa Monica, CA 90407-2138
1200 South Hayes Street, Arlington, VA 22202-5050
201 North Craig Street, Suite 102, Pittsburgh, PA 15213
RAND URL: http://www.rand.org/
To order RAND documents or to obtain additional information, contact Distribution Services: Telephone: (310) 451-7002; Fax: (310) 451-6915; Internet: order@rand.org

PREFACE

Despite the diverse array of civil service occupations, all General Schedule (GS) personnel are paid according to a commonly structured pay table. Critics charge that this common pay table hampers personnel management flexibility in that managers cannot easily compete with the variety of external market opportunities that are available to civil service workers in different occupations. Furthermore, longevity increases are nearly automatic, and promotions can be vacancy driven and given only to eligible individuals, so civil service managers have few methods to provide financial incentives to attract, retain, and motivate higher-quality personnel. Consequently, the critics conclude that the civil service compensation and personnel systems do not adequately compensate superior performance or provide sufficient inducement for higher-quality personnel to stay.

One of the few tools potentially available to personnel managers to reward better performance is accelerated promotion, which results in pay growth. But whether promotion speed varies much across occupational areas and whether better-quality personnel are indeed promoted faster, are paid more, and stay longer are open questions.

The research presented in this report addresses these questions. It uses data on GS civil service workers in the Department of Defense (DoD) to describe career profiles (i.e., the promotion, pay, and retention profiles of groups of personnel) and to estimate whether higher-quality workers are promoted faster, are paid more, and remain longer in DoD civil service. It also provides some evidence on whether these profiles and results have changed in recent years since the DoD drawdown changed the nature of civilian careers in the organization.

This study was sponsored by the Office of Civilian Personnel Policy of the Under Secretary of Defense for Personnel and Readiness. It was prepared within the Forces and Resources Policy Center of RAND's National Defense Research Institute, a federally funded research and development center sponsored by the Office of the Secretary of Defense, the Joint Staff, the unified commands, and the defense agencies.

CONTENTS

FIGURES

TABLES

White-collar General Schedule (GS) workers in federal civil service jobs are covered by a commonly structured pay table that varies somewhat to account for differences in federal and nonfederal pay across geographic areas. Although civil service personnel managers can use special pays and other forms of compensation to help attract, retain, and motivate high-quality employees, some critics have argued that the common pay table hampers the managers' ability to manage flexibly the large and extraordinarily diverse federal GS workforce. Specifically, the critics charge that pay cannot be readily varied in such a way as to motivate higher-quality workers to enter and stay in the civil service.

While they must adhere to the commonly structured pay table, personnel managers do have other tools that enable them to offer higher lifetime pay to higher-quality workers. Most notably, they can promote higher-quality workers faster. However, the degree to which individuals can be promoted quickly depends on whether they are eligible for promotion and on whether vacancies exist in higher grades in specific locations. Furthermore, even if higher-quality personnel are promoted faster and thereby receive higher earnings over their careers, whether these increases are enough to induce them to stay in the civil service is an open question.

Relatively little analysis has been conducted on the career outcomes of higher-quality personnel in the federal civil service or on whether these employees are paid more, are promoted faster, or stay longer than lower-quality personnel do. The research presented in this report begins to fill that gap by examining the pay, promotion, and retention profiles of civil service workers in the Department of Defense (DoD), the largest employer of GS personnel in the federal government. The analysis uses a longitudinal database constructed from Defense Manpower Data Center (DMDC) personnel files. The database tracks the careers through fiscal year 1996 (FY96) of individuals who entered or reentered the DoD civil service between FY82 and FY96. The analysis focuses on those who entered or reentered in FY88, before the defense drawdown, and those who entered or reentered in FY92, during the drawdown. These groups are called the FY88 and FY92 cohorts, respectively. The analysis of the two groups allows comparisons of the results for GS personnel who entered before the drawdown with results for those who entered during it, to determine whether higher-quality personnel are paid more, are promoted faster, and stay longer in the DoD. Regression models are estimated that control for observable characteristics other than personnel quality that might affect career outcomes but that are not related to personnel quality.

Three measures of personnel quality are used in the analysis: supervisor rating, level of education on entering the DoD, and (for the analysis of retention) promotion speed. Each measure has its advantages and its drawbacks. For example, entry education captures general skill level and possibly general ability level. However, education level is measured with error in the DMDC dataset, which could result in biased estimates of whether better-educated employees are paid more, are promoted faster, and stay longer. However, except in the pay analysis, which uses education in each year of service as a covariate, only entry education level is used, and there is no reason to believe that the measurement error for entry education is not random. Supervisor rating captures how well-suited individuals are for the civil service and how well they perform in their jobs, from their supervisors' perspectives; however, ratings are often missing, especially for individuals' first year of employment, and they exhibit relatively little variation across personnel. Promotion speed also captures how well-matched individuals are with their civil service jobs, and it potentially shows more variation across personnel. However, promotion speed is observed only for those who stay and are promoted, and it can be influenced by the vacancy rates in upper pay grades. Because no single measure is perfect in the DMDC data, all three measures are used, and more weight is placed on the overall direction of the results than on specific magnitudes.

Analysis of data for both the FY88 and FY92 cohorts indicates that higher-quality personnel are generally paid more and are promoted faster than lower-quality personnel, regardless of which measure of quality is used and despite the drawbacks of the measure. Specifically, those who receive better ratings from their supervisors are found to earn more and to be promoted faster. The results also indicate that those with any college education are promoted faster and are paid more than those who have not attended college. However, although those with the highest degrees, i.e., master's degrees or doctorates, are found to generally be paid more, when other observable characteristics are held constant, the results indicate that they are not always promoted faster than those with only a bachelor's degree. Thus, having advanced degrees does not seem to always translate into faster promotions, although having any college education seems to do so. Problems with the measurement of the education variable in the pay analysis could cause the effects of education on pay to be underestimated. Nonetheless, the results suggest that the civil service's compensation and personnel systems are generally successful at promoting higher-quality personnel faster and paying them more.

The analysis of earnings and promotion speed also indicates that these outcomes vary considerably across occupational areas in the DoD, even when other observable job and individual characteristics are held constant. Thus, despite the one-size-fits-all pay table that is shared by personnel in all occupations, pay and promotion outcomes vary significantly.

A key question is whether the faster promotion speed and higher pay that higher-quality personnel receive is sufficient to induce them to stay longer in the DoD civil service. The results of the analysis reported here are not overwhelmingly positive. The answer depends on the quality measure, the cohort, and the other variables included in the analysis.

Those in the FY88 cohort who received better ratings from their supervisors had better retention in the DoD, as did those who were promoted faster. Those in the FY92 cohort who were promoted faster also had better retention. However, better-educated personnel in the FY88 cohort, particularly those with advanced degrees, had poorer retention. Among the FY92 cohort, only those with a bachelor's degree were found to stay longer in the DoD, while those with more advanced degrees had poorer retention. Thus, while some of the evidence suggests that better performers in the DoD civil service have better retention, other evidence, especially that based on education, does not. As noted earlier, it appears that personnel with advanced degrees are not always promoted faster than those with bachelor's degrees. The evidence on the retention of personnel with advanced degrees indicates either that they did not fit well in the civil service or that their slower promotion speed translated into poorer retention.

The analysis also indicates that retention patterns vary significantly by occupational area, even when other observable characteristics are held constant. Surprisingly, scientists and engineers, a large percentage of whom have advanced degrees, are found to be among the groups with the best retention, when other factors such as education are held constant. However, whether this retention is sufficient to meet current and future personnel requirements is another open question.

The evidence presented in this report suggests that higher-quality GS personnel in the DoD civil service are generally paid more, are promoted faster, and sometimes are retained longer. It also indicates areas where retention and promotion problems may exist, specifically among the most educated personnel, i.e., those with advanced degrees. However, because of measurement error, the results pertaining to education are less than rock solid. The analysis indicates large variations in the careers of personnel in different occupations, despite the common pay table that serves them. Given the varying requirements for personnel across occupations and the variety of external market opportunities that exist in different occupations, the differences in the careers of GS personnel are no doubt in part a result of these variations.

Future research should examine whether the retirement system that covers federal civil service personnel induces higher- or lower-quality personnel to retire. As the civil service workforce ages along with the rest of the U.S. population in the next few decades, the issue of whether higher-quality personnel hasten or defer retirement will gain in importance. Future research should also seek to refine and define new measures of personnel quality. Such measures might include actual measures of worker productivity or test scores on job-relevant skills. Another issue that should be addressed in future research is the role of bonuses. Although few employees get bonuses, the number is growing, and bonuses can be sizable. Therefore they may play an important role in the retention of higher-quality personnel. Finally, future research should examine whether the career outcomes examined in this study are sufficient to attract and retain a workforce that meets current and future civil service personnel requirements. As more is understood about civil service personnel and compensation systems and the effects of these systems on personnel outcomes, more cost-effective policies can be developed to manage the federal civil service workforce in the years ahead.

ACKNOWLEDGMENTS

This report benefited from the highly capable assistance of Rachel Louie, at RAND, who constructed the longitudinal data files used in the analysis. These data files were based on personnel files maintained by the Defense Manpower Data Center (DMDC). At DMDC, Amanda Marty, Debbie Eittleberg, and Mike Dove provided the data we needed and answered our questions. I would like to thank Jim Hosek and Sue Hosek, also at RAND, for their valuable comments on the research as it progressed, and Al Robbert for his outstanding support as project leader and his useful comments on the research. I would also like to thank Steve Haider, at RAND, and Craig College, of the Office of the Army Chief of Staff, who provided excellent review comments and input. I also benefited from the comments of Larry Lacy, the project monitor. Finally, I am grateful for the input and support of the project sponsor, Diane Disney, Deputy Assistant Secretary of Defense for Civilian Personnel Policy.

Chapter One

INTRODUCTION

Critics of the personnel and compensation systems in the federal civil service charge that these systems are inflexible and hamper the ability of civil service managers to attract, retain, and motivate higher-quality personnel (Kettl, Ingraham, Sanders, and Horner, 1996; Committee on Scientists and Engineers in the Federal Civil Service, 1993; Johnston, 1988). Since managers are limited in their ability to reward better performers, the systems are thought to provide few if any financial incentives for performance. A particular target of this criticism is the common General Schedule (GS) pay table. Step increases, whereby individuals within a pay grade get pay increases based on their longevity, are nearly automatic. Personnel managers are able to reward through faster promotion workers who perform better only if the workers are eligible for promotion and vacancies exist in a higher grade.

Furthermore, the pay table applies to all GS workers, despite the vast array of occupations in which they work and the variety of external market opportunities that may be available to them. Although the GS pay table does vary by geographic location to account for differences in federal and nonfederal pay across areas, it does not vary by occupational area. Critics charge that the one-size-fits-all pay table prevents personnel managers from competing with the private-sector opportunities available to individuals in different occupations, and they conclude that the civil service compensation system is not set up to attract, retain, and motivate higher-quality personnel.[1]

---

[1]Indeed, such criticisms helped spur the creation in the 1980s and 1990s of several demonstration projects that sought, among other things, to circumvent the GS pay table and provide specific agencies with more flexible personnel and compensation

Despite the apparent rigidity of the GS pay table, managers do have some tools at their disposal to compensate and retain those who perform better or who have better private-sector opportunities. One such tool is job security. The civil service is widely perceived as offering significantly more job security than private-sector employment does. To the extent that higher-quality personnel desire job security, the civil service may be a relatively more attractive option for them. However, job security may also be a perquisite for lower-quality personnel, in which case it may not be a relatively stronger incentive for better employees to enter or stay in the civil service. Furthermore, because of the dramatic declines in the size of the civilian workforce in the Department of Defense (DoD) that resulted from to the post–Cold War defense downsizing in the 1990s, job security is less of a perquisite for the workforce as a whole than it was historically.

Another tool for rewarding higher-quality workers is the bonus. Civil service personnel managers have the authority to use hiring, retention, and relocation bonuses to attract and retain civil service employees. A recent study of these payments (U.S. Office of Personnel Management, 1999) found that they have been concentrated in specific occupations (e.g., computer specialties) and in specific areas (e.g., Washington D.C.). Personnel managers who were surveyed in that study indicated that bonuses have been effective overall in retaining personnel in these areas. The study also found that the bonuses or allowances can be quite large, averaging $8,200 in 1998. Nonetheless, bonuses may not be as effective as they could be. Managers who were surveyed indicated that the maximum bonus that could be paid was not high enough, and although the use of such bonuses is growing, less than 1 percent of the civil service employees in the Executive branches of the federal government get them. In addition, while these bonuses might be effective retention tools, whether they are effective in retaining *high-quality* employees is an open question.

A third tool for force-management flexibility in the civil service is promotion speed. When vacancies are available, managers can promote better performers more rapidly and consequently can provide them with higher lifetime civil service earnings. Since promotion

---

systems. The results of one of these experiments, Pacer Share, is described in Orvis, Hosek, and Mattock (1993).

speed can be varied across and even within occupational areas, it can also be used to circumvent the one-size-fits-all pay table. In addition, insofar as the actions of those in the upper grades in a hierarchical organization have a positive effect on the productivity of those in the lower grades, it is important that higher-quality personnel in the lower grades be identified and promoted to the upper grades faster than lower-quality personnel. However, as discussed later, promotion speed also has some problems as an indicator of quality.

Moreover, promotion speed is effective as a financial incentive only if there are vacancies in the upper grades. The defense drawdown, high-grade constraints, and the aging of the baby boom generation have resulted in a more senior civilian workforce in the DoD and fewer promotion opportunities for junior workers. Consequently, the retention of higher-quality younger civilians in the DoD may have declined in recent years.

Little is known about the variation in promotion speed among GS workers in the civil service—whether higher-quality personnel are promoted faster and are paid more, or whether those who are promoted faster and who are of higher quality stay longer in the civil service. Nor is much known about whether higher-quality personnel have been more likely to stay since the beginning of the DoD drawdown. A recent RAND study (Gibbs, forthcoming) examines the extent to which higher-quality scientists and engineers in the DoD laboratories are promoted and retained but does not consider non-laboratory personnel and personnel in other civil service occupational areas.

The study reported here begins to fill that gap by describing variations in promotion speed, retention, and pay among recent cohorts of DoD civil service personnel; by developing proxy measures of personnel quality; and by using those measures to examine whether higher-quality GS workers are promoted faster and retained longer in the DoD civil service than lower-quality workers are, as well as whether these patterns have changed in recent years.

To conduct this analysis, we constructed an unusual longitudinal database on GS workers in the DoD that tracks through fiscal year 1996 (FY96) the careers of those workers who entered or reentered the DoD civil service between FY82 and FY96. The database uses personnel records maintained by the Defense Manpower Data Center (DMDC). The data include information on entry characteristics and how they vary over each individual's career, pay levels, promotion events and timing, and the timing of exits from the DoD civil service. By tracking the careers of several cohorts, the database permits a comparison of careers across occupational groups and time periods, including a period covering part of the DoD drawdown.

The report is organized as follows. Chapter Two presents a simple framework for understanding how the quality measures used in this analysis relate to the personnel outcomes. Chapter Three describes the empirical methodology used to analyze whether higher-quality personnel are paid more, whether they experience faster promotion, and whether they stay in the DoD civil service longer than lower-quality personnel. It also provides a brief overview of the database and its construction, describes the measures that are used to indicate personnel quality, and summarizes the advantages and disadvantages of those measures. Chapter Four presents the first set of analytical results and describes the variations in career profiles— promotion, retention, and pay—across occupational areas for the FY88 and FY92 cohorts of personnel. The variation in the profiles provides some indication of the extent to which outcomes differ across occupational area and the extent to which occupational areas may be managed differently. Chapter Five presents the second set of analytical results and provides estimates of whether higher-quality personnel are promoted faster, are paid more, and stay longer. Chapter Six summarizes the findings and discusses potential areas for future research.

Chapter Two

ANALYTICAL FRAMEWORK

The analysis presented in this report relies on three measures of personnel quality—education, supervisor rating, and promotion speed—and focuses on three personnel outcomes—pay, promotion speed, and length of stay. This chapter describes the hypothesized relationships between the quality measures and the outcome measures. The empirical implementation and the data used are described in Chapter Three. This chapter provides a simple, brief overarching framework and a context for the results that are presented later.

The analytical framework can be represented by a series of equations, the first of which specifies the factors that affect personnel quality. It is hypothesized that personnel quality at time t, Q_t, captures factors that make the individual productive both within the civil service at his or her specific job and outside the civil service. These factors include education and training, innate ability and talent, and motivation. Individual productivity also depends on the individual's job, the skill requirements of the job, the type of equipment used, and the relationship of the job to the organization's overall output. More formally, Q_t is given by

$$Q_t = Q(\text{education, motivation, ability, job factors}) \qquad (2.1)$$

Few data exist on the factors that determine personnel quality. We use entry education as a quality measure, but as Equation 2.1 makes clear, education is only one of many factors.

Quality affects the way supervisors assess the productivity of civil service personnel. The accuracy of supervisor ratings can be affected by other factors as well, including how frequently the supervisor can monitor the worker's output, the method used to monitor output, whether individual output can be easily observed or can be observed only at significant cost, and the supervisor's subjective bias. More formally, the supervisor rating at time t, A_t, is given by

$$A_t = A(Q_t, \text{monitoring frequency, technology, and cost,} \\ \text{subjective assessment}) \qquad (2.2)$$

When a vacancy exists in a non-entry-level position, whether it is filled by promoting from within or by hiring externally depends on the supervisor's assessment of qualified and available personnel in lower-level grades. Therefore, speed of promotion to a higher grade depends on job vacancy rates, the individual's willingness to accept the responsibilities associated with working at a higher grade, and the supervisor's assessment, including previous ratings. More formally, speed of promotion, P, is given by

$$P = P(A_t, \text{job vacancy, willingness to move up, eligibility} \\ \text{for promotion}) \qquad (2.3)$$

Promotion speed in Equation 2.3 depends on supervisor rating, A_t, which, in turn, depends on personnel quality, Q_t. The approach to estimating Equation 2.3 is discussed in Chapter Three. Because supervisor rating captures not only quality but also other factors relating to monitoring and the accuracy of the supervisor's assessment, both education—a determinant of Q_t—and supervisor rating are included as covariates in the analysis of promotion speed. With education included as a covariate, supervisor rating captures the effect on promotion speed of other determinants of Q_t as well as factors that affect the accuracy of the rating.

Whether higher-quality personnel decide to stay in the civil service or leave depends on an array of factors. While developing a model of retention in the civil service is beyond the scope of this analysis, these factors will reflect the individual's internal and external opportunities and the determinants of these opportunities. The opportunities are captured by such variables as pay, benefits, and promotion speed, the individual's taste for federal service or nonfederal service, health status, and job flexibility. More formally, retention, R_t, is given by

$$R_t = R(P, \text{ pay and benefits inside and outside the civil service,}$$
$$\text{taste for federal service)} \tag{2.4}$$

As will be discussed in more detail in later chapters, the retention equation is estimated in two ways. The first excludes P (promotion speed) and includes supervisor rating and education, which may affect both the internal and external opportunities of an individual. For example, a better-educated individual may be able to get a faster promotion within the DoD as well as in an external job. If external opportunities are relatively more attractive, the effects of education and supervisor rating on retention will be negative. The second estimation method includes P. If promotion speed reflects internal opportunities, including P as a covariate in the regression equation means that education and supervisor rating capture the effects of external opportunities on retention. That is, when promotion speed is included, the effects of these variables on retention are hypothesized to be negative.

The final outcome variable examined in this analysis is earnings at time t, represented by S_t. Earnings are determined by a pay table and exclude bonuses in this analysis. Earnings obviously depend on grade and seniority, since these factors affect an individual's placement in the pay table. However, other factors as well affect pay levels and the rate at which pay grows over time. For example, as indicated in Equation 2.3, promotion speed, which affects pay growth, depends on supervisor rating A_t, which in turn depends on Q_t. Pay also depends on labor market experience, seniority within the civil service, occupation, and various job characteristics. More formally, S_t is given by

$$S_t = S(A_t, \text{ occupation, experience, seniority)} \tag{2.5}$$

As discussed in more detail in Chapter Three, Equation 2.5 is estimated using both A_t and education as covariates. These factors are both used because supervisor assessment, A_t, captures quality factors other than education that affect civil service pay.

EMPIRICAL METHODS AND DATA

This chapter describes the empirical methods used in this analysis and the dataset that was constructed for it. The dataset is a longitudinal file that tracks through FY96 the civil service careers of GS personnel who entered or reentered the DoD civil service between FY82 and FY96. "Careers" means the promotion, pay, and retention profiles of these workers, as well as their entry job and individual characteristics and how these characteristics have changed over time. The methods used include estimation of ordinary-least-squares regression models to analyze pay and estimation of Cox regression models to analyze times to promotion and to separation. Since a key purpose of the analysis is to determine whether higher-quality personnel are promoted and retained, the measures used to indicate personnel quality, along with their advantages and their drawbacks, are also described.

EMPIRICAL METHODS

Analysis of Pay

To analyze pay profiles in the DoD civil service and whether higher-quality personnel are paid more than lower-quality personnel, the cohort data are configured so that each observation in the data file corresponds to an individual/year-of-service combination. Workers with more years of service will have more observations in the dataset. The following regression model is then estimated for each cohort:

$$\ln(S_{it}) = \alpha + \beta X_{it} + \gamma_t E_{it} + \nu_t t_{it} + \varepsilon_{it} \qquad (3.1)$$

where S_{it} is annual earnings, i defines the individual, and t defines the year of service (YOS). X_{it} represents a vector of individual and job characteristics. The variables of interest, the quality measures (described later in this chapter), are included in X_{it}. The estimator β represents the effects of X_{it} on earnings in the DoD civil service. Thus, the βs will provide an estimate of how pay varies for higher-quality personnel, holding other observable characteristics constant. Positive βs will provide evidence that higher-quality personnel are paid more.

E_{it} is a set of variables that indicate the individual's years of service at entry. These variables control for entry experience for the roughly 40 percent of each cohort that entered with more than one month of recorded federal service. The effects of these variables on pay are to be estimated and are denoted as γ_t. t_{it} is a set of variables that indicate the individual's years of service since entry. The effects of these variables on pay are estimated and are denoted as v_t. The results concerning how pay varies in the civil service with years of service and how it varies by occupational area focus on v_t.[1] The variable ε_{it} represents the error term in the regression model. It is assumed to be normally distributed, with a mean of zero and standard deviation equal to σ. Equation 3.1 is estimated by ordinary least squares.

A potential bias that arises in the estimation of Equation 3.1 with ordinary-least-squares methods is that annual earnings are observed only for those who stay in service. If those who leave are of lower quality (or higher quality) and would have been paid less (or more) than those who stay, pay by years of service will be underestimated (or overestimated) if the model is not corrected for selection bias.

Various approaches can be used to test and correct for selection bias. The following approach is taken in this analysis: For each year of service, denoted t, the cohort data are divided into two groups: (1) those who separate at year t and (2) those who stay beyond year t. Equation 2.1 is estimated for each group separately. The estimates based on group 1 data are for those who separate, while the estimates based on group 2 data are for those who do not. Thus, the estimates provide a lower and upper bound at each year of service of

---

[1]The results by occupational area are based on estimating a separate Equation 2.1 for each occupational area.

how pay grows with year of service through year t. If the estimates for the two groups do not differ much, selection bias does not appear to be much of a problem. Chapter Five reports the results when t = 8 for the FY88 cohort and t = 4 for the FY92 cohort.[2]

Another potential bias is created by the exclusion of bonuses and special pays from earnings, the dependent variable. If better-quality workers are more likely to get bonuses and special pays, the coefficient estimates for the covariates measuring personnel quality will be biased downward. While this is not a perfect solution, the analysis controls for factors such as occupational area and geographic region that may partially explain why some individuals get bonuses and special pays. If the bias persists despite these controls, then the reported estimates of whether higher-quality personnel are paid more are conservative. That is, the true estimates are larger than those reported.

A final bias is created by measurement error in the education variable, one of the measures of personnel quality and a covariate in the pay analysis. If the measurement error is greater for better-educated workers—and such would be the case if the measurement error is due to problems in consistently updating the education data for those who obtain more education while in the civil service—then the estimated effect on pay of having more education will be biased downward. Measurement error is more problematic in the pay analysis than in the analysis of promotion speed and retention because the pay analysis uses education in each year of service as a covariate. The promotion and retention analyses use only education at entry, a variable that is less likely to be subject to the measurement problem.

---

[2]The results shown in Chapter Four focus on the final year of service for each cohort, YOS 8 for the FY88 cohort and YOS 4 for the FY92 cohort. However, to show the upper and lower bounds of the pay profile for each cohort, it is necessary to estimate the pay regression for both those who stay and those who leave at every year of service, i.e., for YOS 1, YOS 2, YOS 3, . . . through YOS 8 for the FY88 cohort and through YOS 4 for the FY92 cohort, not just for the final year of service. That is, two pay regressions should be estimated for every possible leaving year, not only for the final year for each cohort, in order to control for the possibility that those who leave at an earlier year of service have different unobservable quality characteristics than from those of the individuals who leave later. We estimated such regressions, but they are not shown here because the qualitative results are similar to those found using only the final year of service, and the amount of regressions results is enormous.

Analysis of Promotion Speed and Retention

Analysis of promotion speed and time until separation from DoD civil service (i.e., retention) requires empirical techniques that allow for the examination of the occurrence and timing of "events" and the factors that influence them. In the case of promotion speed, the event is promotion, and in the case of time until separation, the event is separation. The technique used is survival analysis, specifically, the Cox proportional-hazard-model approach with time-varying covariates.

The standard approach for analyzing retention and promotion is to estimate binary-choice regression models that focus on estimating the factors that influence the probability of retention or promotion. There are two primary advantages to using survival analysis techniques rather than more conventional techniques. First, survival analysis accounts for both the occurrence *and* the timing of promotion and separation, whereas conventional methods focus on just the occurrence of these events. But timing can be important. For example, if nearly everyone in a cohort is promoted at least once but they vary in their timing of promotion, conventional methods will detect little variation in the outcome variable (promotion), while survival methods will detect the variation in timing. Furthermore, survival analysis allows for the possibility that some covariates—for example, an individual's supervisor rating—may change over time. Survival methods permit variables in the model to change over time, while conventional methods do not permit variables to change over time.

Second, survival methods account for "censoring." Censoring occurs when the data end before the event occurs. For example, in the case of separation, an individual in the FY88 cohort may not have separated before FY96, when the data end. While the individual will separate from the civil service eventually, the separation is not observed in the data. In the case of promotion, censoring might occur because either the individual has not been promoted before FY96 or the individual separated before being promoted. In the former case, the individual might have been promoted after FY96, but this is not observed in the dataset. Accounting for censoring is important because large numbers of observations may be censored and serious biases in the estimates may result. Censoring is not easily handled with conventional methods, but with survival methods, it is straightforward.

In survival analysis, the time until an event occurs is assumed to be the realization of a random process. The hazard function or hazard rate is used to describe the probability distribution of event times. The hazard function is defined as the risk of the event occurring in month t + 1, given that it did not occur in month t. Formally, the hazard function, h(t), is

$$h(t) = f(t)/S(t)$$
$$\text{with } S(t) = Pr\{T > t\} \text{ and } f(t) = dS(t)/dt \tag{3.2}$$

where S(t) is the cumulative survival function. It gives the cumulative probability that the event time T is greater than t. For example, it indicates the cumulative probability that an individual is promoted after month t. f(t) is the probability-density function. The hazard function is used to describe the probability-distribution function in survival analysis because it can be interpreted as the probability an event occurs at time t given it did not occur at t − 1. In the case of promotion, it indicates the probability of promotion in a month for those who have not been promoted before that month.

In the Cox proportional-hazard model with time-varying covariates, the hazard function is given by:[3]

$$h_i(t) = \lambda(t)\exp(\beta_1 X_i + \beta_2 Z_i(t)) \tag{3.3}$$

Or,

$$\ln h_i(t) = \alpha(t) + \beta_1 X_i + \beta_2 Z_i(t) \tag{3.4}$$

where $\lambda(t)$ is the baseline hazard function, or the hazard for individuals whose covariates X_i and Z_i are zero. X_i is a vector of time-

---

[3]Unlike the better-known and more widely used Cox proportional-hazard model with time-invariate covariates, where the hazard for any individual is a fixed proportion of the hazard for any other individual, the models estimated in this study do not assume proportionality; they incorporate time-dependent covariates $Z_i(t)$, which cause the proportionality assumption to be violated. However, like the proportional model, the nonproportional model can be estimated using partial-likelihood techniques, which produce coefficient estimates without the need to specify the baseline hazard function.

independent covariates, while $Z_i(t)$ is a vector of covariates that vary with time.

In the case of promotion, the specific model estimated is

$$\ln p_i(t) = \alpha(t) + \beta_1 X_i + \beta_2 Z_i(t) \qquad (3.5)$$

where $p_i(t)$ is the hazard of promotion for individual i, and X_i is a vector of job and individual characteristics that are measured at entry and that do not vary with time. These characteristics include race and ethnicity, gender, entry geographic region, entry grade, and months of service at entry. $Z_i(t)$ is a vector of individual characteristics that vary with time, i.e., the cumulative number of years for which the individual has a supervisor rating and the cumulative number of years for which he or she received each rating level (1 to 5).

The data and the variables included in X_i and $Z_i(t)$ are described in more detail later in this chapter, but it should be noted here that two important sets of variables included in X_i and $Z_i(t)$ are the measures of personnel quality. Since a key personnel goal of the civil service is to identify, develop, and promote better-quality personnel, it is hypothesized that the higher-quality personnel will be promoted faster in the DoD civil service. The results presented in Chapter Five provide some evidence on whether or not the data support this hypothesis.

The vectors of coefficients β_1 and β_2 are estimated using partial-likelihood techniques. The baseline hazard $\alpha(t)$ is not estimated. Equation 3.5 is estimated for both the first promotion and the second promotion. In the first case, promotion speed is defined as months in service until first promotion, while in the second case, it is defined as months in current grade until second promotion. The coefficient estimates indicate the estimated effect of each covariate on the promotion hazard, or monthly promotion rate. A positive coefficient means that the covariate is estimated to increase promotion speed.

In the case of separation, the specific model that is estimated is:

$$\ln r_i(t) = \pi(t) + \delta_1 X_i + \delta_2 Z_i(t) \qquad (3.6)$$

where $r_i(t)$ is the separation hazard for individual i, and $Z_i(t)$ is defined the same as in the promotion model. X_i includes the same variables as the promotion model does but with the addition of promotion time. That is, X_i includes a set of variables indicating months in service until first, second, third, and fourth promotion. In the cases of censored observations, the promotion variables are set to the month of separation.[4] The vectors of coefficients δ_1 and δ_2 are estimated using partial-likelihood methods. The coefficient estimates indicate the estimated effect of each covariate on the separation hazard, or monthly separation rate. A positive coefficient means that the covariate is estimated to increase the speed of separation, i.e., reduce retention.

A key set of variables included in X_i and $Z(t)$ in the retention analysis is the set of quality measures described later in this chapter. While a full theory of the determinants of retention in the civil service is beyond the scope of this analysis, previous analyses have shown that the effect of personnel quality on retention is ambiguous and cannot be predicted *a priori*. That is, theory cannot predict whether higher-quality personnel are more likely to stay in the civil service or less likely (Asch and Warner, 1994; Buddin, Levy, Hanley, and Waldman, 1992). The reason is that higher-quality personnel have better opportunities than lower-quality personnel, both inside and outside the civil service. Whether higher-quality personnel are more likely to stay depends on whether the incentives to stay are stronger than the incentives to take advantage of good external market opportunities. This issue is examined empirically in Chapter Four. Specifically, the analysis examines whether the δ_1 and δ_2 are positive, indicating that higher-quality personnel have a higher separation hazard, or negative, indicating that they have a lower separation hazard (i.e., they stay longer).

The coefficient estimates, together with their standard errors, permit the computation of a Wald statistic for each estimate. The Wald statistic has a chi-squared distribution. The statistical significance of

---

[4]An alternative specification is estimated in which promotion speed is represented by a variable equal to the cumulative number of promotions received at each month t, until the censor point. Estimation of this alternative specification yields results quite similar to those reported in Chapter Four.

each estimate is determined by whether the Wald statistic has a probability greater than 5 percent.

Chapter Five presents the results of estimating Equations 3.1, 3.5, and 3.6. The longitudinal data files used to estimate these equations are described below.

DATA

The longitudinal data files were created from the beginning-fiscal-year inventories and transaction files for DoD civil service personnel from FY82 to FY96, obtained from DMDC. The inventory data include every permanent GS employee in the DoD civil service as of the beginning of each fiscal year, and the transaction data indicate changes in each individual's personnel record during the year, including appointments and reappointments, promotions, and separations.

The longitudinal data track the careers of those who entered or reentered the DoD civil service in each fiscal year. Each fiscal year defines a cohort—for example, those who entered or reentered in FY88 define the FY88 cohort. The longitudinal dataset tracks the careers of the cohort over the eight-year period from FY88 to FY96. Individuals who were in the DoD civil service before FY88 and who were not reentrants in FY88 are not included in the FY88 cohort. Since many of those in each cohort in the dataset are reentrants, the months of service at entry can vary from 0 for new hires up to 360 (30 years).

The data do not permit easy differentiation between new civil service entrants and rehires. The months-of-service variable in the DMDC data includes months of active-duty military service, so a new entrant to the civil service may have more than zero months of service, indicating that he or she is a veteran. In the FY88 cohort data, 16 percent of those who had prior military service entered the civil service with months of service greater than zero.

Some individuals appear to be rehires because their months-of-service variable at entry is greater than zero, but they are not veterans. About 22 percent of the FY88 cohort fall into this category. Not all of these apparent rehires appear in earlier inventory data, however, suggesting that they may actually be new entrants with potentially miscoded months-of-service data or new hires to the DoD who previously worked in another federal agency.[5] The analysis does not try to distinguish new entrants from rehires. The modeling approach simply accounts for whether an individual enters with months of service greater than zero and whether he or she is a veteran by including controls for prior military service and for months of service or years of service at entry.

Some individuals are observed to have a gap in their DoD civil service. That is, they are observed to enter, stay for a period of time, leave, and then return. About 5 percent of the FY88 cohort fall into this category. While empirical methods to account for gaps in service exist, these individuals have been excluded from this analysis[6]; this exclusion does not affect the results qualitatively.

Some individuals are excluded from each cohort file because they were temporary workers, worked less than full-time, were considered "inactive," were seasonal, or were military technicians. They have been excluded because they may be less attached to the workforce than the typical permanent GS worker, and they may follow different career tracks.

A key set of variables in the analysis comprises months until each promotion and months until separation from the DoD. These variables were developed by scanning both the inventory and trans-

---

[5]In the FY88 cohort, 58 percent of the individuals entered with months of service equal to zero and are not veterans. These individuals are clearly new entrants. About 4 percent of the FY88 cohort entered as veterans with months of service equal to zero. About 30 percent of these individuals were under the age of 40 and unlikely to be military retirees; therefore, they should have gotten credit for their military experience. The months-of-service variable is probably miscoded in these cases. These observations were deleted from both the FY88 and FY92 cohorts for the purpose of the analysis. Veterans who entered with months of service equal to zero but who were over 40 years of age are potentially military retirees who opted to not accept credit for their military service. They are included in the analysis.

[6]To account for the possibility that the outcomes differed for individuals who had a gap in their service, a separate analysis (not shown here) was performed to include them in the dataset, and a variable was included in the regression analysis to indicate whether the individual had a gap. The results regarding the pay and promotion and, to some extent, the retention of high-quality personnel did not differ much from the results presented in this report. However, convergence was not always achieved in the estimation of the partial-likelihood function of the time-until-separation models. Because their inclusion would have produced incomplete results, these observations were excluded in the analysis presented here.

action files to determine whether a promotion or separation took place. If so, the date and therefore the number of months until the promotion or separation occurred are shown in the transaction files. If a transaction record was not available but a promotion or separation seemed to have occurred, according to the inventory data,[7] then the last available record—either the inventory or the last transaction record—was used to determine separation or promotion date. In the cases where the data are "censored," i.e., they end either because the files end at FY96 or because the individuals separated from service, the months-to-promotion variable is set to the months until the data are censored. Similarly, the months-until-separation variable is set to months of service until the separation occurred or until the data are censored for those who were still in service in FY96. It should be noted that the data indicate separation from the DoD, not necessarily from the civil service. Some individuals who leave the DoD transfer to other agencies within the federal civil service. Although the data do provide some indication of whether a separation from the DoD is a transfer rather than an exit, the data on transfers are incomplete, according to DMDC. On the other hand, separations from the DoD are more clearly observable. Therefore, the analysis simply uses months until separation from the DoD rather than separation from the civil service.

Another key variable is annual earnings. Annual earnings include base pay only and not other pays and bonuses the individual might have received. It is defined as gross earnings, not net earnings. Data on earnings are taken from the annual inventory record for each individual. If the individual separated, the last recorded annual-earnings information is used for the last year of service. All earnings are translated into constant 1996 dollars, using the annual Consumer Price Index for urban residents.

Data in the longitudinal data files include a wide range of information on each individual, including job and individual characteristics.

---

[7]It should be noted that when there is no transaction record indicating that a separation from the civil service occurred but the individual disappears from the inventory data, it is possible that the individual left the DoD civil service but transferred to another federal civil service agency and therefore remained in the federal civil service. For these cases, separation is defined as separation from DoD civil service.

The files are constructed so that the data indicate both entry characteristics and how these characteristics change over the individual's career. Job characteristics include occupational area, component (Army, Navy, Air Force, Navy, or other defense agency), entry pay grade, months of federal service, last supervisor rating, and supervisor or managerial status. The specific occupations included in the occupational areas are shown in the Appendix. Individual characteristics include gender, race and ethnicity, education, geographic region, veteran's status, retirement system coverage, reported handicap status, and age.

Two important data-quality problems should be noted. First, when the inventories and transaction files were consecutively strung together by social security number[8] to track each individual's career over time, the months-in-service variable did not always increment in a sensible fashion. The records of these individuals were deleted from the data file, as were those of individuals who had a gap in their service over the data period.[9] A second data problem relates to the education variable. According to DMDC, the education variable in the civilian personnel files is not always accurately recorded, nor is it accurately and consistently updated as individuals accumulate more education. Since one of the measures of personnel quality used in this analysis is education (see the discussion later), this is particularly troublesome. This problem was addressed by using more than one measure of personnel quality in the analysis and not relying exclusively on education. Furthermore, only *entry* education is included in the regressions for the promotion and retention analysis. Therefore, consistent updating of the variable was not required. Nonetheless, if entry education is mismeasured, the coefficient estimates on the entry-education variables could be biased. On the other hand, if the measurement error is random, it should not be a problem for this analysis.

---

[8]Social security numbers in the DMDC data were scrambled by DMDC before the data files were sent to RAND to protect the confidentiality of the individuals represented in the file. Since the scrambled social security number is consistent across years for each individual, we were able to match the annual records.

[9]The problem of nonsequential years-of-service data in the DMDC files is discussed in more detail in Asch and Warner (1999), Appendix B.

Although longitudinal files were constructed for every fiscal year entry cohort from FY82 to FY96, results are presented in the following chapters for only the FY88 and FY92 cohorts, for brevity. Earlier cohorts were not used because information was missing for several key variables, including supervisor rating, one of the quality measures used. The FY88 cohort represents a group of individuals who entered or reentered prior to the DoD downsizing, while the FY92 cohort captures individuals who entered or reentered during the downsizing. Although the FY88 cohort entered prior to the drawdown, part of the careers of this cohort span the drawdown years after FY90. Thus, the FY88 cohort gives a partial picture of the predrawdown promotion and retention experiences of DoD civil service personnel.

Figure 3.1 shows a schematic of the longitudinal data files that were constructed and used for the analysis. Individuals in each cohort enter and then are promoted over their career. For the FY88 cohort, the

RAND*MR1193-3.1*

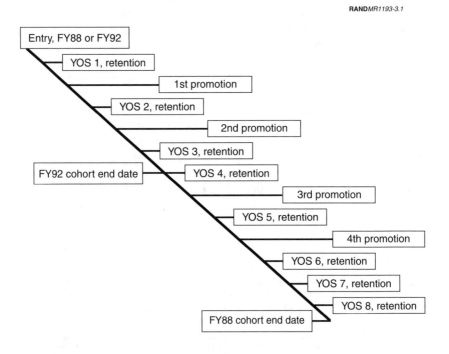

Figure 3.1—Schematic Diagram of the Data Configuration for Each Cohort

data end after eight years, while for the FY92 cohort, the data end after four years. Over the course of the data period, individuals are observed making retention decisions and leaving or staying in the civil service. For visual simplicity, these decisions are shown to occur annually in the figure, although in the data analysis, the decisions are assumed to occur monthly.

Table 3.1 presents some summary statistics of the entry characteristics of the FY88 and FY92 cohorts. It also gives the variable names and definitions used in the analysis.

Table 3.1

Variable Means for the FY88 and FY92 Cohorts

Variable	Definition	FY88 Cohort (N=31,912)	FY92 Cohort (N=19,744)
NONWHITE	Race/ethnicity is non-white, non-Hispanic	0.270	0.290
AGE21_0	Age between 21 and 30 at entry	0.499	0.476
AGE 31_0	Age between 31 and 40 at entry	0.285	0.286
AGE 41_0	Age between 41 and 50 at entry	0.137	0.151
AGE 51_0	Age between 51 and 60 at entry	0.042	0.046
AGE 61_0	Age 61 or older at entry	0.006	0.008
HCAPCAT0	Reported disability	0.059	0.008
HCAPMIS0	Reported disability missing	.014	0.006
FEMALE	Female	.578	0.605
DMDCVET	Prior military service (DMDC definition)	0.169	0.203
BELOWHS_0	Education = below high school degree	0.007	0.008
HSG_0	Education = high school degree at entry	0.310	0.418
SOMECOL_0	Education = some college at entry	0.280	0.209
AADEG_0	Education = AA degree at entry	0.051	0.045
BADEG_0	Education = BA degree at entry	0.266	0.233
ABOVBA_0	Education = Above BA degree/professional degree (no MA)	0.025	0.019
MA_0	Education = MA degree at entry	0.051	0.058
PHD_0	Education = PhD at entry	0.009	0.010
MNYOS0	Months in service at entry	27.385	31.714
EGRADE	Entry grade	5.426	5.650
RAT1_0	Rating =1 (highest supervisor rating)	0.012	0.031
RAT2_0	Rating = 2	0.026	0.033
RAT3_0	Rating = 3	0.118	0.036
RAT4_0	Rating = 4	0.000	0.001
RAT5_0	Rating = 5 (lowest supervisor rating)	0.001	0.000
RATMIS_0	Rating is missing	0.832	0.890
PROF_0	Professional	0.205	0.183
ADMIN_0	Administrative	0.113	0.118
TECH_0	Technical	0.143	0.271
CLERICAL_0	Clerical	0.469	0.355
BLUECOL_0	Blue collar	0.000	0.000

Table 3.1 (continued)

Variable	Definition	FY88 Cohort (N=31,912)	FY92 Cohort (N=19,744)
WHITCOL_0	Other white collar	0.071	0.074
ARMY_0	Army	0.379	0.429
NAVY_0	Navy	0.345	0.216
MARINES_0	Marine Corps	0.017	0.018
AIRFORCE_0	Air Force	0.142	0.171
OTHAG_0	Other defense agency	0.117	0.165
OPMMIS_0	Region = missing/foreign	0.134	0.204
S_EAST_0	Region = Southeast	0.112	0.114
G_LAKES_0	Region = Great Lakes	0.081	0.112
MID_CON_0	Region = Mid-continental	0.036	0.025
NEWENG_0	Region = New England	0.053	0.035
EASTERN_0	Region = Eastern	0.049	0.043
MID_ATL_0	Region = Mid-Atlantic	0.245	0.191
ROCKIES_0	Region = Rockies	0.025	0.029
S_WEST_0	Region = Southwest	0.061	0.082
WESTERN_0	Region = Western	0.172	0.131
N_WEST_0	Region = Northwest	0.030	0.036
FM10_0	Occupation =science	0.026	0.024
FM11_0	Occupation = engineering	0.119	0.081
FM11_0	Occupation = engineering	0.119	0.081
FM20_0	Occupation = mathematics	0.004	0.003
FM21_0	Occupation = medical	0.025	0.038
FM22_0	Occupation = legal	0.000	0.000
FM23_0	Occupation = education	0.000	0.000
FM24_0	Occupation = misc. professional	0.021	0.032
FM30_0	Occupation = logistics managemt	0.036	0.032
FM31_0	Occupation = personnel managemt	0.000	0.000
FM32_0	Occupation = financial managemt	0.022	0.024
FM33_0	Occupation = data systems management	0.022	0.018
FM34_0	Occupation = central managemt	0.030	0.033
FM40_0	Occupation = science and engineering technician	0.035	0.022
FM41_0	Occupation = medical technician	0.016	0.043
FM42_0	Occupation = logistics technician	0.000	0.000
FM43_0	Occupation = management technician	0.033	0.041
FM44_0	Occupation = miscellaneous technician	0.091	0.209
FM50_0	Occupation = secretary	0.255	0.111
FM51_0	Occupation = financial clerk	0.029	0.049
FM52_0	Occupation = logistics clerk	0.047	0.036
FM53_0	Occupation = general office operations	0.062	0.059
FM54_0	Occupation = misc. clerical	0.071	0.116
FM60_0	Occupation = medical attendants	0.000	0.000
FM61_0	Occupation = Fire/police	0.054	0.031
FM62_0	Occupation = personnel services	0.000	0.000

Perhaps the most dramatic difference between the two cohorts is in their size. Because of downsizing in the DoD civilian workforce, especially in the early 1990s, there were 38 percent fewer new entrants in FY92 than in FY88. Other notable differences are in the occupational mix and the agency mix of the GS entrants and in their ages. The FY92 cohort had significantly fewer clerical workers, but far more technical workers. It also had far fewer Navy workers, but more employed in the various defense agencies and in the Air Force. The fraction who were veterans rose from about 17 percent to 20 percent, most likely reflecting the relatively larger pool of individuals with prior military service in FY92 as a result of the decrease in the size of the active-duty military force and the existence of veterans-preference policies in the federal civil service. The percentage entering over the age of 50 rose in the FY92 cohort, as did the percentage with a high school diploma.

MEASURES OF PERSONNEL QUALITY

As discussed in Chapter Two, the analysis summarized in this report uses three measures of personnel quality. The first two, education and supervisor ratings, are used to analyze whether higher-quality personnel experience more pay growth, are promoted faster, and are retained longer. The third measure, promotion speed, is studied as both an indicator of personnel quality and as a personnel outcome. As a personnel quality indicator, it is used in the analysis of whether higher-quality personnel are retained. All three measures have advantages and drawbacks. For that reason, more than one measure is used in the analyses.

The first measure, education, captures the individual's general skill level. That is, it captures skills that could be used in both the civil service and other job opportunities. In addition to the advantage of capturing general skill levels, entry educational level is also easily measurable. Furthermore, because civil service jobs seldom have hard-and-fast degree requirements, observed promotion rates of better-educated workers into higher grades is not an artifact of existing job requirements, but reflects the supply of better-educated workers to higher grades. Unfortunately, as noted above, the data for this variable are subject to measurement error. Moreover, although those who have more education may be better-quality workers in general, they are not necessarily better in the civil service, especially if civil service employment requires specialized skills. Some individuals may not be well suited to the civil service or may not be well matched to their civil service jobs. The quality of their job performance may, in fact, be lower because of their educational level. Put differently, educational level does not capture the quality of the job match.

The second measure of personnel quality used in this study is supervisor rating. This measure addresses one of the concerns about the usefulness of the education variable in that it indicates the quality of a worker's performance from the supervisor's standpoint and therefore provides an indication of how well-suited the individual is for his or her civil service job. However, supervisor rating has two drawbacks as a measure of personnel quality. First, it is sometimes missing in the DMDC files, especially for the first year of service. Table 3.1 shows that supervisor rating is missing for more than 80 percent of the entrants in the first year. By the second year, the number with missing values drops significantly to around 15 percent. To address the problem of missing values, the estimated earnings regressions (shown in Chapter Five) include a variable that indicates whether the rating variable is missing for a particular year of service. In the Cox regressions of promotion and retention, a variable is included that indicates the cumulative number of years for which the variable is missing. By incorporating variables indicating a missing supervisor rating, the analysis accounts for the possibility that those who have a missing value for the supervisor rating are more likely to be promoted, retained, or paid more.

A second drawback of supervisor rating as a measure of quality is that it has limited variance. A supervisor rating can take only one of five values, with 1 being the best and 5 being the worst rating.[10] Furthermore, the vast majority of GS personnel receive a rating of

_____

[10]The ratings are: 1 = outstanding; 2 = exceeds fully successful; 3 = fully successful; 4 = minimally successful; 5 = unsatisfactory. During the 1990s, some individuals were under a "pass/fail" scale rather than the "five-step" scale. Unfortunately, the scales could not be differentiated in the data. Consequently, estimates in this report for supervisor rating capture the effects of both scales.

1, 2, or 3, further limiting the variance of this measure.[11] In the regressions that include rating as a covariate, indicator variables identify individuals who got a rating of 1 or 2, while the rating in the omitted categories is 3, 4, or 5. A final problem with supervisor rating as a measure of performance is that it can be subject to "supervisor bias," whereby some supervisors give consistently better ratings than others. Furthermore, insofar as supervisors cannot perfectly monitor or observe all of a worker's activities, supervisor ratings may not provide a complete picture of a worker's performance level. Because of these problems, a third measure of personnel quality, promotion speed, is used.

Promotion speed is measured in months. Like supervisor rating, months until promotion can provide information about the quality of the match between the individual and the civil service. Those who are promoted faster are presumably best suited and the best performers among those eligible for promotion. Using promotion speed as a measure of personnel quality also addresses one of the concerns regarding supervisor rating, namely, limited variance. Promotion speed may vary significantly across individuals and can potentially make wider and finer quality distinctions among them than can supervisor rating. Comparison of the retention patterns of those who are promoted faster relative to the patterns of those who are promoted slower can also provide some indication of whether higher-quality personnel are retained.

However, promotion speed is also problematic as an indicator of personnel quality. If those who are chosen for promotion are simply those who are the most willing to stay in the civil service, faster promotion speed will not necessarily identify the superior performers. Instead, it will indicate taste for civil service or some other unobserved characteristic that makes some workers "stayers" and some "leavers." That is, it may not be the case that fast promotees are retained; it may be that those who are likely to be retained are promoted. This bias is not addressed in the analysis. Consequently, we cannot conclude causation, but only correlation between promotion speed and retention. Another potential problem with promotion

---

[11]In the FY88 inventory of GS personnel, fewer than 1 percent of the workers had received a supervisor rating worse than 3 in their past assessment.

speed is censored data. If few individuals are promoted prior to separation or prior to the date when the data end, promotion speed will be missing for most individuals and will not provide much indication of quality. Fortunately, a nontrivial number of individuals in both the FY88 and FY92 cohorts were observed to experience at least one promotion, although the figure is larger in the FY88 cohort than in the FY92 cohort because the data extend for a longer period of time and because promotion rates were lower overall during the drawdown period in the early 1990s. Table 3.2 shows the fraction who get a first, second, third, and fourth promotion in each cohort.

Another potential problem with promotion speed as an indicator of quality is that it may simply reflect vacancy rates in the civil service. If some parts of the civil service consistently experience poorer or better retention and have more or fewer vacancies in the senior grades than others, promotion speeds in those areas may reflect differential vacancy rates rather than differential personnel quality. Therefore, the analysis of promotion controls for occupational area, geographic region, and other observed job and individual characteristics that may give rise to different vacancy rates across the civil service.

Because none of the quality measures is ideal and all are subject to problems, all three measures are used here. If results are consistent for all three measures, this lends credence to their overall validity. Furthermore, in the conclusions drawn from the analyses, more weight is given to the overall direction of the estimated effects—i.e., whether higher-quality personnel are retained—than to the specific magnitudes of the estimated effects.

Table 3.2

Percentage Receiving Promotions in the FY88 and FY92 Cohorts

	FY88	FY92
First promotion	71.8	43.0
Second promotion	37.9	20.6
Third promotion	26.6	10.1
Fourth promotion	16.4	1.7

Chapter Four

CAREER OUTCOMES BY OCCUPATIONAL AREA

This chapter presents some background information on how occupational areas differ in terms of average entry grade, promotion speed, retention, and pay. This background information sheds some light on the degree to which occupations vary in personnel outcomes. While the results are not conclusive, especially since no control is included in these comparisons for other factors that may affect the outcomes observed, significant variation would suggest that personnel managers may have some flexibility in how fast individuals are promoted, whether they are paid more, and, therefore, whether they are retained longer. The specific occupations associated with each occupational area are listed in the Appendix.

MEAN ENTRY GRADE

Characteristics at entry, particularly entry pay grade, are important descriptors of the career profiles of cohorts of personnel. Table 2.1 showed the entry characteristics, including mean entry grades, for the FY88 and FY92 cohorts. The mean entry grades for the two cohorts were 5.4 and 5.7, respectively. Figure 4.1 shows that mean entry grade varied considerably across occupational areas in both cohorts, while the differences between the two cohorts were much smaller.

As one would expect, occupational areas where entrants have more education and therefore better external market opportunities have a higher mean entry grade. This is to be expected because civil service managers must offer higher pay in order to compete successfully with the private sector for better-educated workers and workers in

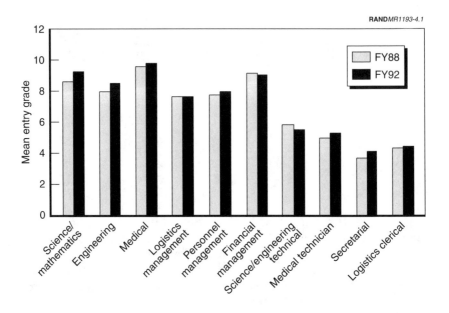

Figure 4.1—Mean Entry Grade of Permanent GS Workers in Selected Occupational Areas, FY88 and FY92 Cohorts

technical areas. Those in science and mathematics, engineering, and the medical and financial-management fields have higher mean entry grades than those in the clerical and technician areas. For example, workers in science and mathematics entered in FY92 at a mean grade of 9.2, while secretarial workers entered at an average grade of 4.0.

PROMOTION PROFILES

While one might expect entry grades to be higher in occupational areas where private-sector opportunities are better, it is unclear *a priori* how promotion profiles should vary by occupational area. A complete theory of the determinants of faster promotion in the civil service by occupational area is beyond the scope of this analysis. Still, it is likely that those determinants will include the retention and therefore the available vacancies in the upper grades, personnel quality, and the transferability of skills from the civil service to the external market.

Figure 4.2 shows the cumulative probability curve, S(t), defined in Equation 3.2, for months to first promotion for the FY88 and FY92 cohorts. To compute the cumulative promotion probability we grouped the data into three-month intervals. The figure shows the cumulative probability at the midpoint of each interval. Although the FY88 cohort extends for a longer period, the curve stops at month 60 (the point where the FY92 data end) to enable comparison of the profiles across cohorts. For ease of illustration, the figure shows the curves for only a sample of the occupational areas.

In the first three-month interval, no one in either cohort was promoted. By month 24 (i.e., two years), part of each cohort had been promoted. By month 60, even more of the cohorts had been pro-

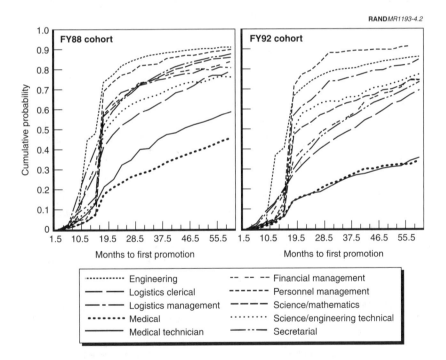

Figure 4.2—Cumulative Probability of First Promotion,
FY88 and FY92 Cohorts

moted. Although the cumulative probabilities increase with month, they increase at a decreasing rate, i.e., the cumulative probability curve is concave with respect to the origin.

Figure 4.2 shows considerable variation in speed to first promotion across occupational areas. Engineers and workers in personnel management receive the fastest promotions. By 60 months (five years), about 90 percent of the workers in these areas in both cohorts had been promoted at least once. In contrast, by 60 months, about 70 to 75 percent of those in logistical clerical occupational areas had achieved their first promotion.[1] Promotion in the medical and medical technician occupational areas was even slower. In the FY92 cohort, only 35 percent of employees in these area had been promoted after five years. As shown in Figure 4.1, workers in the medical area enter the civil service at a higher grade, but they do not achieve promotions as often as those in other occupational areas. Clearly, medical workers follow a different career track from that of workers in other occupational areas.

Figure 4.2 also shows that except in a few occupational areas, workers in the FY92 cohort were promoted at a somewhat slower pace than those in the FY88 cohort. For example, by six months, 45 percent of the engineers in the FY88 cohort had been promoted, as contrasted to only 38 percent of those in the FY92 cohort. Promotion was also substantially slower for the medical occupations in the FY92 cohort. In the FY88 cohort, 60 percent of the medical workers had achieved their first promotion by month 60, but only 35 percent in the FY92 cohort had done so.

Figure 4.3 shows similar patterns in the cumulative survival curves for speed to second promotion for the FY88 and FY92 cohorts. Months to second promotion vary considerably across occupational areas, with engineers and workers in personnel management receiving the fastest promotions. Workers in the FY88 cohort achieved

---

[1]The differences in speed to first and second promotion across occupational areas (see Figures 4.2 and 4.3) are generally statistically significant. Statistical significance is determined by the significance of the coefficient estimates on the occupational indicator variables in the estimated Cox regression models for promotion speed, shown in Tables 5.2 and 5.3. These coefficient estimates are generally statistically significant at the 5 percent level, even after observed job and individual characteristics are controlled for.

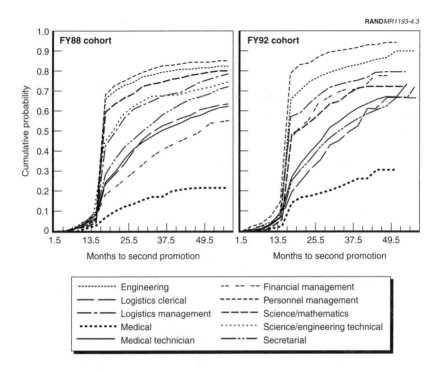

**Figure 4.3—Cumulative Probability of Second Promotion,
FY88 and FY92 Cohorts**

their second promotions somewhat faster than those in the FY92
cohort.

RETENTION PROFILES

Retention profiles also vary by occupational area. Models of retention behavior predict that the decision to stay in the DoD civil service
will be affected by individual tastes and the expected financial payoff
associated with civil service employment relative to that in the external market.[2] Although a complete model of retention is beyond the
scope of this analysis, it is clear that if tastes and the financial payoffs

---

[2]See Asch and Warner (1994) for a model of the decision to stay in service for active-
duty military personnel.

associated with civilian employment vary by occupational area, retention profiles will vary by occupational area as well.

Figure 4.4 shows the survival, or cumulative probability curves, for months until separation by cohort for a selected set of occupational areas. At the beginning of the first month, everyone in the entering cohort was in the civil service, implying that the cumulative probability of staying in service is 1. After two years (24 months), between 55 and 90 percent of the workers were still in service, depending on occupational area and cohort. Engineers and those in logistics management stay the longest, or are the least likely to leave in each cohort. About 70 percent of individuals in those occupational areas are still in service after five years. In contrast, about 35 percent of the medical technicians in the FY88 cohort and about 30 percent of those in the FY92 cohort were still in service after five years.

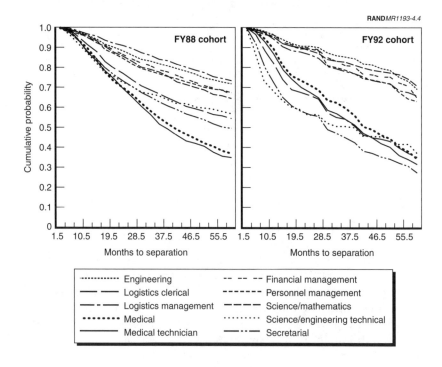

Figure 4.4—Cumulative Probability of Retention, FY88 and FY92 Cohorts

PAY

Given the differences in promotion speed across occupational areas, one would expect pay growth to differ by occupational area as well. Since all occupational areas use the common pay table, the primary means by which real pay growth can vary is promotion speed. To examine differences in pay growth, controlling for observable job and individual characteristics, we estimated Equation 3.1 for each occupational area, using the FY88 cohort data. The resulting variation in pay growth by years of service is shown in Figure 4.5.[3]

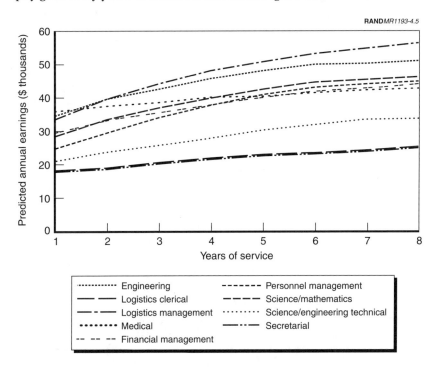

Figure 4.5—Pay Profiles by Occupational Area, with Job and Individual Characteristics Controlled For, FY88 Cohort

---

[3]As discussed in Chapter Three, the cohort data were divided into two groups to estimate the pay profiles for each year of service. The results shown in Figure 4.5 are for personnel who stayed beyond YOS 8. Also, the estimates were converted from a log scale to a linear scale to make the results more readily accessible.

The differences in the pay profiles across occupational areas shown in Figure 4.5 are dominated somewhat by the differences in mean entry pay shown in Figure 4.1. That is, the biggest differences in the profiles appear to be the differences in the intercepts or in the relative height of the profiles. The height of the profile is determined by the mean entry grade for each occupational area. To control for entry grade and to focus only on pay growth, Figure 4.6 shows the percentage difference between pay at YOS 1 and pay at YOS 8, with observed characteristics held constant in the regression framework. For example, between YOS 1 and YOS 8, real pay grew by 50 percent in the science and mathematics area, by about 40 percent for engineers, and by about 65 percent for workers in personnel management, with observed individual and job characteristics controlled for.

The results in Figure 4.6 show that real pay grew considerably over the eight-year period for the FY88 cohort. Furthermore, pay growth varied across occupational areas. Those in the medical field experi-

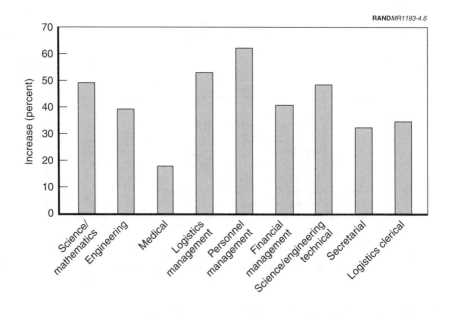

Figure 4.6—Percentage Increase in Annual Earnings from YOS 1 to YOS 8
by Occupational Area, with Job and Individual Characteristics
Controlled For, FY88 Cohort

enced the least pay growth, while those in personnel management experienced the most. Given that promotion speed is fastest in personnel management and slowest in the medical field, these results are not surprising.

Chapter Five

PAY, PROMOTION, AND RETENTION
OF HIGHER-QUALITY PERSONNEL

This chapter presents the results of the analysis of personnel quality, using the personnel-quality measures education, supervisor rating, and promotion speed to examine whether higher-quality personnel in the DoD civil service are paid more, are promoted faster, and stay longer.

PAY

To examine whether higher-quality personnel have higher salaries, we estimated ordinary-least-squares regression equations for two groups of personnel in each year of service, those who stay until that year and leave and those who stay beyond that year. Table 5.1 shows the estimation results for those in the FY88 cohort who stayed until or beyond YOS 8 and those in the FY92 cohort who stayed until or beyond YOS 4. As discussed in Chapter Three, the pay regression was estimated for both groups as a means of addressing the potential problem of selection bias caused by the fact that annual salary is observed only for those who stay. The estimates for the two groups provide a lower and an upper bound at YOS 8.

The results in Table 5.1 indicate the estimated percentage change in annual earnings as a result of an increase in the variables. In the case of RATING1 (the highest supervisor rating an individual can get), the estimate indicates the percentage change in earnings when the rating is 1 relative to being 3, 4, or 5 (the omitted categories). For the FY88 cohort, an "outstanding" rating increased annual earnings by

Table 5.1
Ordinary-Least-Squares Regression Results, by Entry Cohort
(Dependent Variable = Log(Annual Salary))

| | FY88 COHORT | | | | FY92 COHORT | | | |
| | Beyond YOS 8 | | Left at YOS 8 | | Beyond YOS 4 | | Left at YOS 4 | |
	Estimate	Std. Error	Estimate	Std. Error	Estimate	Std. Error	Estimate	Std. Error
INTERCEP	9.755*	0.004	9.744*	0.015	9.553*	0.009	9.376*	0.028
DMDCVET	-0.008*	0.002	-0.021*	0.006	0.019*	0.004	0.048*	0.015
RATING1	0.028*	0.001	0.043*	0.005	0.038*	0.004	0.022	0.012
RATING2	0.023*	0.001	0.038*	0.004	0.027*	0.003	0.018	0.011
RATMIS	-0.008*	0.002	-0.019*	0.005	0.000*	0.004	-0.006	0.010
EDMIS	0.135*	0.028	0.141	0.175			0.207**	0.081
SOMECOL	0.017*	0.001	0.008	0.004	0.006*	0.003	0.021**	0.009
AADEG	0.022*	0.002	0.011	0.008	0.050*	0.005	0.047*	0.017
BADEG	0.077*	0.002	0.077*	0.006	0.062*	0.004	0.011	0.011
ABOVBA	0.107*	0.003	0.077*	0.011	0.081*	0.007	0.030	0.026
MA	0.073*	0.002	0.058*	0.009	0.056*	0.005	0.068*	0.020
PHD	0.116*	0.005	0.069*	0.019	0.078*	0.010	0.064	0.045
NONWHITE	-0.014*	0.001	-0.007	0.004	-0.024*	0.003	0.002	0.007
FEMALE	-0.017*	0.001	-0.012*	0.004	-0.033*	0.003	-0.036*	0.009
HCAPMIS	0.009	0.016	-0.013	0.042	0.093**	0.035	0.016	0.075
HCAPCAT	-0.017*	0.002	-0.021*	0.006	-0.006*	0.005	0.031	0.016
SYOS2[a]	-0.059*	0.002	-0.029*	0.008	-0.072*	0.005	-0.068*	0.014
SYOS3[a]	-0.075*	0.002	-0.072*	0.009	-0.085*	0.006	-0.138*	0.022
SYOS4[a]	-0.125*	0.003	-0.104*	0.008	-0.122*	0.008	-0.174*	0.036
SYOS5[a]	-0.169*	0.003	-0.135	0.009				
SYOS6[a]	-0.179*	0.004	-0.147	0.012				
SYOS7[a]	-0.199*	0.005	-0.186	0.016				

Table 5.1 (continued)

| | FY88 COHORT | | | | FY92 COHORT | | | |
| | Beyond YOS 8 | | Left at YOS 8 | | Beyond YOS 4 | | Left at YOS 4 | |
	Estimate	Std. Error	Estimate	Std. Error	Estimate	Std. Error	Estimate	Std. Error
YOS2	0.092*	0.002	0.061*	0.007	0.075*	0.004	0.044*	0.009
YOS3	0.150*	0.002	0.107*	0.007	0.118*	0.004	0.076*	0.011
YOS4	0.213*	0.002	0.157*	0.007	0.163*	0.004	0.121*	0.023
YOS5	0.260*	0.002	0.181	0.007				
YOS6	0.299*	0.002	0.209	0.007				
YOS7	0.318*	0.002	0.217	0.008				
YOS8	0.337*	0.002	0.231	0.013				
AGE20	-0.007	0.005	-0.001*	0.014	-0.085*	0.008	-0.092*	0.021
AGE30	0.009*	0.002	0.031*	0.006	0.012*	0.004	-0.001	0.016
AGE40	-0.008*	0.001	0.007*	0.005	0.003*	0.004	-0.023	0.015
AGE50	-0.003	0.002	0.023**	0.006	0.005*	0.004	0.006	0.017
AGE60	-0.020*	0.002	0.017*	0.007	-0.010*	0.005	-0.006	0.022
EGRADE4	0.074*	0.001	0.086*	0.005	0.120*	0.004	0.153*	0.009
EGRADE 5	0.176*	0.002	0.167*	0.006	0.226*	0.004	0.224*	0.012
EGRADE 6	0.199*	0.004	0.172*	0.012	0.307*	0.008	0.319*	0.023
EGRADE 7	0.302*	0.002	0.301*	0.007	0.412*	0.005	0.427*	0.016
EGRADE 9	0.387*	0.002	0.392*	0.009	0.496*	0.006	0.457*	0.020
EGRADE 11	0.483*	0.003	0.521*	0.009	0.642*	0.006	0.647*	0.023
EGRADE 12	0.624*	0.003	0.676*	0.013	0.771*	0.007	0.818*	0.030
EGRADE 13	0.804*	0.008	0.965*	0.039	0.990**	0.013	0.917*	0.054
EGRADE 14	0.966*	0.013	1.059*	0.064	1.166**	0.019	1.104*	0.066
EGRADE 15					1.313**	0.030	1.402*	0.127
FM10	0.142*	0.003	0.166*	0.013	0.174*	0.007	0.205*	0.033
FM11	0.316*	0.002	0.337*	0.008	0.341*	0.005	0.372*	0.021

Table 5.1 (continued)

| | FY88 COHORT | | | | FY92 COHORT | | | |
| | Beyond YOS 8 | | Left at YOS 8 | | Beyond YOS 4 | | Left at YOS 4 | |
	Estimate	Std. Error	Estimate	Std. Error	Estimate	Std. Error	Estimate	Std. Error
FM21	0.141*	0.005	0.168*	0.013	0.249*	0.008	0.211*	0.022
FM24	0.147*	0.004	0.178*	0.013	0.145*	0.007	0.174*	0.024
FM30	0.103*	0.003	0.091*	0.010	0.129*	0.007	0.103*	0.028
FM32	0.172*	0.003	0.190*	0.014	0.192*	0.007	0.232*	0.028
FM33	0.153*	0.003	0.160*	0.012	0.164*	0.008	0.185*	0.032
FM34	0.109*	0.003	0.127*	0.011	0.109*	0.007	0.125*	0.028
FM40	0.075*	0.003	0.106*	0.011	0.047*	0.009	-0.099*	0.034
FM41	-0.089*	0.005	-0.052*	0.013	0.050*	0.007	0.025	0.019
FM43	-0.070*	0.003	-0.037*	0.009	0.056*	0.006	0.084*	0.020
FM44	0.024*	0.003	0.020**	0.008	-0.010**	0.005	-0.048*	0.012
FM50	-0.070*	0.002	-0.059*	0.007	0.040*	0.006	0.050*	0.014
FM51	-0.090*	0.003	-0.043*	0.010	0.045*	0.006	0.063*	0.019
FM52	-0.106*	0.003	-0.061*	0.008	0.014*	0.008	-0.006	0.023
FM53	-0.101*	0.003	-0.055*	0.008	0.007*	0.006	0.036**	0.017
FM54	-0.117*	0.002	-0.078*	0.008	-0.043*	0.005	-0.099*	0.013
ARMY	-0.017*	0.002	-0.003	0.006	0.063*	0.004	0.175*	0.010
NAVY	0.008*	0.002	0.029*	0.006	0.083*	0.004	0.198*	0.011
MARINE	0.002	0.004	0.016	0.014	0.081*	0.009	0.148*	0.029
AIRFORC	0.004**	0.002	0.008	0.007	0.071*	0.004	0.178*	0.012
COMPET	0.008*	0.002	-0.007	0.005	0.027*	0.003	0.100*	0.009
SUP_MGR	0.064*	0.002	0.060*	0.008	0.070*	0.006	0.114*	0.018
OPMMIS	0.003	0.004	-0.030*	0.011	0.015*	0.007	0.092*	0.019
NEWENG	0.018*	0.003	0.041*	0.011	-0.012*	0.008	-0.020	0.026
EASTERN	0.029*	0.003	0.008	0.011	0.002*	0.007	0.037	0.023

Table 5.1 (continued)

| | FY88 COHORT | | | | FY92 COHORT | | | |
| | Beyond YOS 8 | | Left at YOS 8 | | Beyond YOS 4 | | Left at YOS 4 | |
	Estimate	Std. Error	Estimate	Std. Error	Estimate	Std. Error	Estimate	Std. Error
MID_ATL	0.033*	0.003	0.017	0.010	0.036*	0.006	0.049*	0.019
S_EAST	-0.013*	0.003	-0.021	0.010	-0.002*	0.006	0.036	0.020
G_LAKES	0.014*	0.003	-0.008	0.010	0.036*	0.007	0.061*	0.021
S_WEST	-0.014*	0.003	-0.009	0.011	-0.022*	0.007	0.031	0.021
MID_CONT	-0.011*	0.004	-0.022	0.013	-0.051*	0.009	-0.117*	0.028
ROCKIES	-0.003	0.004	-0.021	0.013	0.008*	0.008	0.091*	0.027
WESTERN	0.023*	0.003	0.024*	0.010	0.018*	0.006	0.031	0.020
N	103,741		11,687		33,629		8,647	
Mean Dep. Variable	10.257		10.158		10.140		9.871	
F-Statistic	7437.688		557.763a		1797.523		170.225	
R-Squared	.843		.783		.785		.578	

Note: * = statistical significance at the 1 percent level; ** = statistical significance at the 5 percent level. See Table 3.1 for definitions of variables.

aSYOS are a set of dummy variables indicating the individuals' years of service at entry.

between 2.8 and 4.3 percent, when other characteristics are controlled for. A rating of 2 ("exceeds fully successful") increased annual earnings for this cohort by a smaller amount, between 2.3 and 3.8 percent. The estimated effects of both variables are statistically significant at the 1 percent level. Thus, those in the top performance categories were paid more than those in the lower performance categories (the omitted group in the regression).

Workers in the FY88 cohort with an associate's degree had between 1.1 and 2.2 percent higher earnings than those with only a high school diploma, other factors held constant. Having less than a two-year college degree, denoted as "some college" in the table, had a very small effect for those who left at YOS 8, between 0.8 percent and 1.7 percent. On the other hand, those who entered with a bachelor's degree had salaries 7.7 to 7.8 percent higher than those who entered with no college (the omitted group). As noted earlier, these estimates might be biased downward because bonuses are omitted from the dependent variable. Thus, the positive relationship between education level and pay may be even greater than is indicated by the table.

The results for the FY92 cohort support the same general conclusions, although the magnitudes of the estimated effects differ somewhat and some effects are not statistically significant at the 5 percent level. Those who had a better supervisor rating and more education were paid more, other factors held constant. An "outstanding" rating increased pay by between 2.2 and 3.8 percent, while an "exceeds fully successful" rating increased pay by between 1.8 and 2.7 percent, with other observed job and individual characteristics controlled for. However, the lower-bound estimates, obtained from the regression results using the "left at YOS 4" group, are not statistically significantly different from zero at the 5 percent level. The upper-bound estimates are statistically significant. An associate's degree at entry is estimated to increase pay by between 4.7 and 5 percent, while a bachelor's degree is estimated to increase pay by between 1.1 and 6.2 percent over the four years of service observed in the data. As in the FY88 cohort, having a doctorate had a large effect on pay; it increased pay by between 6.7 and 7.8 percent over that of workers having no college. Those with a professional degree or other post-baccalaureate education were also paid more in both cohorts; in the FY92 cohort, their pay was between 3 percent and 8.1 percent higher than that of those with no college education, other factors held constant.

The higher salaries of individuals with any college education is consistent with a prevalent economic theory of education and pay growth. Human-capital theory posits that pay is greater among more-educated personnel because those who invest more in education receive skills that are valuable in the job market, they tend to invest more in other productivity-enhancing activities such as informal on-the-job training, and they tend to get more out of these activities, enabling them to earn more in the future. The results in Table 5.1 indicate that better-educated personnel are paid more over their career in the civil service than less-educated personnel, other characteristics held constant.

Table 5.1 also shows the estimated effect on earnings of other job and individual characteristics. Factors that are positively associated with pay include entry pay grade and whether an individual is a supervisor or manager. On the other hand, some factors are negatively associated with pay. Those who enter with more years of service are paid less, all else equal. For example, entering with two years of service reduced pay by between 2.9 and 5.9 percent in the FY88 cohort. This result suggests that when other factors (such as entry grade, occupational area, education, and region) are held fixed, skills other than education that are learned outside the DoD civil service are not generally transferable to the civil service. The effect of prior military service on pay varies by cohort. In the FY88 cohort, having prior military service reduced pay by 0.8 to 2.1 percent, other factors held constant; in the FY92 cohort, those with prior military service received between 1.9 and 4.8 percent higher pay than did those with no prior military service. The effect of service or agency on pay also varied by cohort. Relatively little difference is seen in pay across the armed services in the FY88 cohort. However, in the FY92 cohort, pay was significantly lower in the "other" defense agencies (the excluded category) and higher in the Army, Navy, Air Force, and Marine Corps. The effect of geographic region also varied by cohort. For example, in the FY88 cohort, those in the mid-Atlantic region had 1.7 to 3.3 percent higher pay, but in the FY92 cohort, they had between 3.6 and 4.9 percent higher pay.

The degree to which selection bias is important is demonstrated by the estimated salary profiles by years of service for the FY88 cohort in Figure 5.1 for those who left at YOS 8 and those who stayed beyond YOS 8. The profiles hold variables other than years of service at their

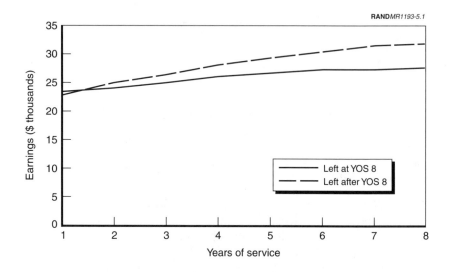

Figure 5.1—Predicted Annual Earnings Profiles, Other Characteristics Controlled For, FY88 Cohort

mean values. For ease of reading, the log scale of the dependent variable is converted to a linear scale. The figure shows that both groups started at about the same level of pay, with observed individual and job characteristics controlled for, but pay grew somewhat faster for those who stayed beyond YOS 8. Specifically, pay started at about $23,000 (in constant FY96 dollars) but grew to about $32,000 by YOS 8 for those who stayed, but to only about $27,000 for those who left. This indicates that those who stay longer earn more over their initial career. To the extent that those who experience faster pay growth are also higher-quality personnel, these results suggest that higher-quality personnel in the FY88 cohort stayed longer. This issue is examined further later in this chapter.

PROMOTION SPEED

Promotion speed is both an outcome and a personnel quality measure. In this section, the focus is on promotion speed as an outcome and on the relationship between promotion speed and the other two measures of personnel quality used in this study, i.e., entry education and supervisor rating.

Table 5.2 shows the results of estimating the Cox regression models of months to first and to second promotion for the FY88 cohort. Table 5.3 shows the results for the FY92 cohort. Of particular interest are the columns labeled Risk Ratio. For indicator variables such as AADEG (associate's degree), the risk ratio, equal to $\exp(\beta)$, can be interpreted as the ratio of the estimated hazard for those with a value of 1 to the estimated hazard for those with a value of 0 (controlling for the other covariates). For example, the estimated risk ratio for AADEG for the FY88 cohort is 1.185, which is greater than 1. This means that the hazard of first promotion is 18.5 percent higher for those with an associate's degree than it is for those with no higher education (i.e., promotion speed is 18.5 percent faster). On the other hand, the risk ratio associated with entering at a pay grade of 4 is 0.664, which is less than 1. This means that the hazard of first promotion is 35.6 percent (1 – 0.644) slower for this group than that for the omitted category (those who enter at a pay grade lower than 4). For quantitative covariates such as CUMRAT1, a more intuitive statistic is obtained by subtracting 1 from the risk ratio and multiplying by 100. This gives the estimated percentage change in the hazard for each one-unit increase in the covariate. For example, an additional year of receiving a supervisor rating equal to 1 increases the hazard by 30.9 percent (1.309 -1 x 100).[1]

The variables CUMRAT1, CUMRAT2, and CHAVRAT are time-varying covariates that capture the effects of supervisor rating on promotion. Because supervisor rating is missing for a significant number of person-years, the variable CHAVRAT indicates the cumulative number of years for which the individual does have a supervisor rating. The variable CUMRAT1 indicates the cumulative fraction of "outstanding" ratings the individual has received, defined as the cumulative number of years for which a rating of 1 has been received divided by the cumulative number of years for which the individual has a supervisor rating in the data. Similarly, the variable CUMRAT2 indicates the cumulative fraction of "exceeds fully successful" ratings the

---

[1]Interpretation of the Cox-regression-model output and a fuller discussion of the different types of survival analytical approaches are given in Allison (1995).

Table 5.2

Partial-Likelihood Cox-Regression-Model Estimates of Months to First and Second Promotion, FY88 Cohort

	First Promotion			Second Promotion		
	Estimate	Std. Error	Risk Ratio	Estimate	Std. Error	Risk Ratio
MNYOS0	0.009*	0.001	1.009	0.023*	0.001	1.023
MNPROM1				-0.023*	0.001	0.977
CUMRAT1	0.269*	0.030	1.309	0.318*	0.034	1.374
CUMRAT2	0.125*	0.024	1.133	0.109*	0.029	1.115
CHAVRAT	0.040**	0.021	1.041	0.122*	0.035	1.130
DMDCVET	0.110*	0.026	1.116	0.033	0.036	1.034
SOMECOL0	0.090*	0.020	1.094	0.065**	0.027	1.067
AADEG0	0.170*	0.037	1.185	0.108**	0.050	1.114
BADEG0	0.415*	0.027	1.515	0.239*	0.036	1.271
ABOVBA0	0.447*	0.053	1.564	0.159**	0.070	1.172
MA0	0.369*	0.044	1.447	0.126**	0.061	1.134
PHD0	0.275*	0.101	1.317	0.381**	0.188	1.464
AGE20_0	1.183*	0.159	3.263	1.236*	0.357	3.441
AGE30_0	1.106*	0.156	3.021	1.054*	0.355	2.870
AGE40_0	0.988*	0.156	2.685	0.881**	0.355	2.413
AGE50_0	0.842*	0.157	2.322	0.657	0.355	1.930
AGE60_0	0.579*	0.160	1.785	0.408	0.360	1.504
RACEMISS	-0.248	0.317	0.780	-0.736	0.583	0.479
NONWHITE	-0.100*	0.018	0.905	-0.112*	0.024	0.894
FEMALE	-0.065*	0.021	0.937	-0.033	0.027	0.967
HCAPMIS0	0.378*	0.069	1.459	0.385*	0.095	1.470
HCAPCAT0	-0.302*	0.037	0.740	-0.231*	0.050	0.794
EGRADE4	-0.440*	0.022	0.644	-0.256*	0.030	0.774
EGRADE5	-0.865*	0.028	0.421	-0.057	0.039	0.944
EGRADE6	-1.029*	0.057	0.357	-0.239*	0.081	0.787
EGRADE7	-0.920*	0.036	0.399	-0.287*	0.047	0.751
EGRADE9	-1.506*	0.042	0.222	-1.383*	0.059	0.251
EGRADE11	-2.316*	0.049	0.099	-2.465*	0.080	0.085
EGRADE12	-2.830*	0.062	0.059	-2.564*	0.124	0.077
EGRADE13	-3.078*	0.154	0.046			
EGRADE14	-2.993*	0.295	0.050			
FM10_0	0.887*	0.066	2.427	1.221*	0.087	3.391
FM11_0	1.368*	0.051	3.928	1.480*	0.071	4.393
FM20_0	0.271*	0.107	1.311	0.005*	0.123	1.005
FM21_0	-0.020	0.088	0.980	-0.410	0.193	0.664
FM24_0	0.842*	0.070	2.322	1.051**	0.098	2.859

Table 5.2 (continued)

	First Promotion			Second Promotion		
	Estimate	Std. Error	Risk Ratio	Estimate	Std. Error	Risk Ratio
FM30_0	0.747*	0.055	2.112	0.811*	0.076	2.250
FM32_0	0.869*	0.070	2.384	1.132*	0.092	3.103
FM33_0	0.995*	0.066	2.704	0.790*	0.095	2.204
FM34_0	0.482*	0.062	1.620	0.718*	0.090	2.050
FM40_0	0.624*	0.053	1.867	0.728*	0.073	2.071
FM41_0	-0.446*	0.087	0.640	-0.345**	0.149	0.708
FM43_0	0.058	0.056	1.059	0.293*	0.078	1.341
FM44_0	0.115*	0.047	1.122	0.475*	0.070	1.609
FM50_0	0.133*	0.041	1.143	0.221*	0.060	1.247
FM51_0	0.205*	0.056	1.228	0.286*	0.079	1.331
FM52_0	-0.018	0.050	0.982	0.159**	0.071	1.172
FM53_0	-0.067	0.048	0.935	0.078	0.069	1.081
FM54_0	-0.167*	0.049	0.846	0.086	0.071	1.090
ARMY0	-0.138*	0.027	0.871	-0.083**	0.035	0.920
NAVY0	0.183*	0.026	1.201	0.094*	0.035	1.098
AIRFORC0	0.209*	0.031	1.233	0.029	0.040	1.029
COMPET0	0.056*	0.023	1.057	0.106*	0.032	1.111
SUPMIS	0.042	0.341	1.043	-0.634	0.720	0.531
SUP_MGR	-0.066	0.062	0.936	0.116	0.090	1.124
OPMMIS0	-0.417*	0.044	0.659	-0.041	0.080	0.960
NEWENG	0.130*	0.045	1.138	-0.097	0.077	0.908
EASTERN	0.165*	0.044	1.179	0.023	0.076	1.024
MID_ATL	0.210*	0.033	1.234	0.139**	0.064	1.150
S_EAST	0.105*	0.036	1.110	-0.089*	0.067	0.915
G_LAKES	0.136*	0.038	1.146	0.177	0.069	1.194
S_WEST	0.011	0.042	1.011	-0.025	0.074	0.975
MID_CONT				0.009	0.078	1.009
ROCKIES	0.176*	0.053	1.193	0.020	0.085	1.021
WESTERN	0.135*	0.035	1.145	0.078	0.066	1.081
N	28,350			17,423		
% censored	35.5			39.5		
-2 log L	367132.2*			169934.5*		

Note: * = statistical significance at the 1 percent level; ** = statistical significance at the 5 percent level. See Table 3.1 for definitions of variables.

Table 5.3

Partial-Likelihood Cox-Regression-Model Estimates of Months to First and Second Promotion, FY92 Cohort

	First Promotion			Second Promotion		
	Estimate	Std. Error	Risk Ratio	Estimate	Std. Error	Risk Ratio
MNYOS0	-0.009*	0.001	0.991	0.027*	0.003	1.027
MNPROM1				-0.055*	0.003	0.946
CUMRAT1	0.297*	0.039	1.345	0.135*	0.049	1.144
CUMRAT2	0.218*	0.039	1.244	0.036	0.047	1.036
CHAVRAT	-0.208*	0.043	0.813	0.065	0.065	1.067
DMDCVET	0.190*	0.037	1.209	-0.008	0.058	0.992
SOMECOL0	0.247*	0.033	1.280	0.138*	0.051	1.148
AADEG0	0.213*	0.060	1.237	0.067	0.093	1.070
BADEG0	0.429*	0.038	1.536	0.258*	0.057	1.294
ABOVBA0	0.454*	0.078	1.574	0.258**	0.107	1.294
MA0	0.457*	0.057	1.579	0.223*	0.086	1.250
PHD0	0.108	0.136	1.114	-0.228	0.392	0.797
AGE20_0	1.060*	0.231	2.887	0.857**	0.365	2.356
AGE30_0	1.027*	0.226	2.794	0.737**	0.358	2.089
AGE40_0	0.890*	0.226	2.434	0.582	0.359	1.789
AGE50_0	0.701*	0.226	2.016	0.407	0.360	1.502
AGE60_0	0.430	0.233	1.538	0.225	0.373	1.253
NONWHITE	-0.124*	0.027	0.883	-0.107*	0.040	0.898
FEMALE	-0.164*	0.028	0.849	-0.165*	0.038	0.848
HCAPMIS0	0.615*	0.152	1.849	-0.082	0.283	0.922
HCAPCAT0	-0.347*	0.056	0.707	-0.148	0.085	0.863
EGRADE4	-0.316*	0.038	0.729	-0.114**	0.057	0.893
EGRADE5	-0.715*	0.044	0.489	0.006	0.066	1.007
EGRADE6	-0.811*	0.078	0.444	-0.568*	0.138	0.566
EGRADE7	-0.711*	0.053	0.491	-0.240*	0.078	0.787
EGRADE9	-1.359*	0.063	0.257	-1.383*	0.099	0.251
EGRADE11	-2.180*	0.072	0.113	-2.902*	0.159	0.055
EGRADE12	-2.974*	0.094	0.051	-2.570*	0.262	0.077
EGRADE13	-3.733*	0.274	0.024			
EGRADE14	-3.085*	0.361	0.046			
FM10_0	1.087*	0.102	2.965	1.191*	0.150	3.292
FM11_0	1.536*	0.085	4.645	1.745*	0.129	5.725
FM20_0	-0.019	0.177	0.981	0.304	0.195	1.355
FM21_0	-0.133	0.120	0.876	-0.073	0.237	0.930
FM24_0	0.849*	0.099	2.336	1.008*	0.152	2.741
FM30_0	0.802*	0.091	2.229	0.835*	0.136	2.304
FM32_0	1.058*	0.096	2.881	1.461*	0.138	4.309
FM33_0	0.921*	0.102	2.511	1.336*	0.148	3.804
FM34_0	0.612*	0.096	1.845	0.583*	0.151	1.792
FM40_0	0.142	0.105	1.152	0.820*	0.148	2.270

Table 5.3 (continued)

	First Promotion			Second Promotion		
	Estimate	Std. Error	Risk Ratio	Estimate	Std. Error	Risk Ratio
FM41_0	-0.957*	0.106	0.384	-0.518**	0.204	0.596
FM43_0	0.384*	0.086	1.469	0.474*	0.131	1.606
FM44_0	-0.082	0.075	0.921	0.352*	0.117	1.422
FM50_0	-0.105	0.080	0.901	0.183	0.125	1.201
FM51_0	0.207*	0.081	1.230	0.253**	0.125	1.287
FM52_0	-0.059	0.094	0.943	0.107	0.151	1.113
FM53_0	-0.204**	0.084	0.815	0.020	0.132	1.020
FM54_0	-0.191**	0.077	0.826	0.366*	0.118	1.442
ARMY0	-0.231*	0.035	0.793	-0.116**	0.052	0.890
NAVY0	0.310*	0.040	1.364	0.096	0.058	1.100
AIRFORC0	-0.071	0.043	0.931	-0.256*	0.066	0.774
COMPET0	-0.125*	0.030	0.882	-0.060	0.047	0.942
SUP_MGR	-0.590*	0.084	0.554	-0.208	0.185	0.812
OPMMIS0	-0.018	0.073	0.982	-0.235*	0.071	0.791
NEWENG	-0.148	0.095	0.862	-0.131	0.108	0.877
EASTERN	-0.075	0.084	0.928	-0.077	0.081	0.926
MID_ATL	0.155**	0.068	1.168			
S_EAST	0.079	0.071	1.083	-0.026	0.057	0.975
G_LAKES	0.203*	0.071	1.225	0.008	0.056	1.008
S_WEST	0.161**	0.074	1.174	0.115	0.065	1.122
MID_CONT	-0.016	0.098	0.984	-0.180	0.118	0.836
ROCKIES	0.221**	0.090	1.247	0.066	0.100	1.068
WESTERN	0.156**	0.072	1.169	-0.029	0.059	0.972
N	16,427			7962		
%censored	51.82			49.76		
-2 log L	123849.5*			2646.42*		

Note: * = statistical significance at the 1 percent level; ** = statistical significance at the 5 percent level. See Table 3.1 for definitions of variables.

individual has received, defined as the cumulative number of years for which a rating of 2 has been received divided by the cumulative number of years for which the individual has a supervisor rating in the data.

An "outstanding" rating is estimated to increase promotion speed substantially, as does having more education at entry, for both the FY88 and FY92 cohorts. In the FY88 cohort, getting another rating of 1 ("outstanding") increased the hazard of first promotion by 30.9 percent and increased that of second promotion by 37.4 percent relative to the excluded group—those who got ratings of 3 ("fully successful"), 4 ("minimally successful"), or 5 ("unsatisfactory"). That is, for an individual who has not been promoted, achieving another "outstanding" supervisor rating increased the probability of getting a promotion in a given month by 30.9 percent for the first promotion and 37.4 percent for the second. For the FY92 cohort, the estimates are 34.5 and 14.4 percent, respectively. Thus, getting the top rating reduced the time to achieve both the first and the second promotion in both cohorts.

Getting another rating of 2, defined as "exceeds fully successful," also increased the promotion hazard, but not as much as getting another rating of 1. In the FY88 cohort, getting another rating equal to 2 increased the hazard of first promotion by 13.3 percent and increased that of second promotion by 11.5 percent. In the FY92 cohort, getting another rating of 2 increased the first-promotion hazard by 24.4 percent and increased the second-promotion hazard by 3.6 percent, although the latter estimate is not statistically significant at the 5 percent level. In summary, those who perform better are estimated to be promoted faster, and the better the performance, the faster is the promotion.

In the FY88 cohort, individuals with a bachelor's degree were estimated to have a hazard of time to first promotion 51.5 percent greater than that of those with no college at entry, and a hazard of time to second promotion 27.1 percent greater. In the FY92 cohort, the estimated effects were 53.6 and 29.4 percent, respectively. The estimated effects of having more than a bachelor's degree were also positive, but not always larger than those of having only a bachelor's degree, and they were not always statistically significant, as one might expect. For example, in the FY88 cohort, having a PhD increased the first promotion hazard by 31.7 percent, which is less than the estimated effect of having a bachelor's degree. The effect of having a PhD degree in the FY92 cohort was not statistically significant.

Tables 5.2 and 5.3 indicate that factors other than the measures of personnel quality influence promotion speed. Those who enter at younger ages experience faster promotions, especially first promotion. Those with prior military service also have faster first promotions, although the pay regression estimates in Table 5.1 indicate that they had somewhat lower pay in the FY88 cohort, other factors held constant. Tables 5.2 and 5.3 also indicate that those who enter at higher grades have slower promotions, although they enter at higher pay levels, as reported in Table 5.1. Consistent with the figures in Chapter Three, promotion speed varies considerably by occupational area, even with other observable characteristics held constant. Those in engineering and science have the fastest promotions, while those in the medical and medical technician fields are estimated to be promoted more slowly. Other factors associated with promotion speed include race, ethnicity, and having a reported handicap.

One factor of note is the relationship between the timing of the first and the second promotion. The positive coefficient estimates on months to first promotion (TPROM1) in the regression model of time to second promotion indicate that those who are promoted more slowly the first time are promoted more slowly the second time. That is, holding other observable factors constant, promotion speed is positively correlated. More specifically, the results indicate that getting a first promotion one month faster increased the monthly hazard rate of second promotion by 2.3 percent in the FY88 cohort and by 5.4 percent in the FY92 cohort, even when some of the factors that affect vacancy rates, such as occupational area and geographic region, are held constant. This suggests that there are "fast-trackers" in the civil service, that is, people who move quickly through the pay table and rise quickly through the organization. To the extent that fast-trackers are higher-quality personnel, this result also suggests that despite the common pay table, the system can reward those who are apparently better performers over time.

RETENTION IN THE DoD

The final outcome examined in this analysis is length of stay until separation from DoD civil service. Table 5.4 shows the results from estimating the Cox regression model of months until separation for the FY88 cohort, and Table 5.5 shows the results for the FY92 cohort. The tables give results for two specifications of the model: The first specification includes all three quality measures (entry education, supervisor rating, and months until each promotion), and the second excludes months until each promotion. Because significantly fewer individuals are observed to have received a promotion in the FY92 data than in the FY88 data (a result of the shorter time period over

Table 5.4

Partial-Likelihood Cox-Regression-Model Estimates of Months to Separation, FY88 Cohort

	Includes Promotion Speed Variables			Excludes Promotion Speed Variables		
	Estimate	Std. Error	Risk Ratio	Estimate	Std. Error	Risk Ratio
CUMRAT1	0.409*	0.032	1.506	-0.603*	0.033	0.547
CUMRAT2	0.347*	0.028	1.415	-0.409*	0.028	0.665
TPROM1	-0.051*	0.001	0.950			
TPROM2	-0.054*	0.001	0.947			
TPROM3	-0.056*	0.001	0.945			
TPROM4	-0.094*	0.002	0.910			
MNYOS0	-0.005*	0.000	0.995	-0.003*	0.000	0.997
CHAVRAT	-0.201*	0.023	0.818	-1.392*	0.024	0.249
DMDCVET	-0.422*	0.025	0.656	0.010	0.025	1.010
EDMIS0	0.351	0.335	1.420	-0.184	0.317	0.832
SOMECOL0	0.013	0.019	1.013	0.014	0.019	1.015
AADEG0	-0.017*	0.038	0.983	-0.086**	0.037	0.918
BADEG0	0.135*	0.027	1.145	-0.020	0.026	0.980
ABOVBA0	0.195*	0.057	1.216	0.035	0.056	1.036
MA0	0.324*	0.045	1.383	0.235*	0.043	1.265
PHD0	0.406*	0.109	1.500	0.257*	0.105	1.293
AGE20_0	0.718*	0.091	2.050	-0.358*	0.087	0.699
AGE30_0	0.443*	0.087	1.558	-0.538*	0.083	0.584
AGE40_0	0.275*	0.087	1.316	-0.786*	0.083	0.456
AGE50_0	0.255*	0.088	1.290	-0.860*	0.084	0.423
AGE6_0	0.236*	0.091	1.266	-0.550*	0.087	0.577
RACEMISS	0.143	0.238	1.154	0.056	0.231	1.058
NONWHITE	0.003	0.018	1.003	-0.120*	0.018	0.887
FEMALE	-0.045**	0.022	0.956	0.111*	0.021	1.117
HCAPMIS0	-0.064	0.074	0.938	-0.083	0.071	0.921
HCAPCAT0	0.085**	0.038	1.089	0.064	0.037	1.067
EGRADE4	-0.203*	0.022	0.816	-0.060*	0.021	0.942
EGRADE5	-0.525*	0.028	0.592	-0.178*	0.027	0.837
EGRADE6	-0.779*	0.053	0.459	-0.167*	0.050	0.846
EGRADE7	-0.351*	0.039	0.704	-0.291*	0.038	0.747
EGRADE9	-0.610*	0.046	0.543	-0.367*	0.043	0.693
EGRADE11	-0.763*	0.054	0.466	-0.364*	0.050	0.695
EGRADE12	-0.899*	0.064	0.407	-0.376*	0.060	0.686
EGRADE13	-0.907*	0.130	0.404	-0.411*	0.129	0.663
EGRADE14	-1.646*	0.218	0.193	-0.923*	0.221	0.397
EGRADE15	-0.504	0.365	0.604			
FM10_0	-0.071	0.078	0.931	-0.650*	0.075	0.522

Table 5.4 (continued)

	Includes Promotion Speed Variables			Excludes Promotion Speed Variables		
	Estimate	Std. Error	Risk Ratio	Estimate	Std. Error	Risk Ratio
FM11_0	-0.077	0.055	0.926	-0.547*	0.052	0.579
FM20_0	-0.223	0.154	0.800	-0.408*	0.151	0.665
FM21_0	0.219*	0.068	1.245	0.517*	0.067	1.677
FM24_0	0.050	0.077	1.052	-0.249*	0.073	0.780
FM30_0	-0.180	0.065	0.835	-0.365*	0.061	0.694
FM32_0	-0.040	0.084	0.961	-0.344*	0.075	0.709
FM33_0	0.139	0.076	1.150	-0.202*	0.072	0.817
FM34_0	-0.051	0.062	0.950	-0.070	0.059	0.933
FM40_0	-0.034	0.056	0.967	-0.048	0.053	0.953
FM41_0	0.104	0.066	1.110	0.357*	0.064	1.430
FM43_0	-0.154*	0.055	0.857	-0.108**	0.054	0.897
FM44_0	-0.188*	0.044	0.829	0.024	0.043	1.024
FM50_0	-0.258*	0.041	0.773	-0.231*	0.040	0.794
FM51_0	-0.194*	0.058	0.824	-0.216*	0.056	0.806
FM52_0	-0.214*	0.050	0.807	-0.226*	0.049	0.797
FM53_0	-0.322*	0.047	0.725	-0.153*	0.045	0.858
FM54_0	-0.188*	0.046	0.829	0.036	0.044	1.036
ARMY0	-0.107*	0.028	0.898	0.182*	0.027	1.199
NAVY0	-0.148*	0.030	0.863	-0.126*	0.028	0.882
MARINE0	-0.046	0.069	0.955	-0.158**	0.065	0.854
AIRFORC0	-0.298*	0.033	0.742	0.063	0.032	1.065
COMPET0	-0.017	0.022	0.984	0.108*	0.021	1.114
SUPMIS	0.248	0.286	1.282	-0.012	0.285	0.988
SUP_MGR	-0.071	0.060	0.932	-0.019	0.058	0.981
OPMMIS0	0.225*	0.053	1.253	0.605*	0.050	1.831
NEWENG	0.318*	0.056	1.374	0.036	0.054	1.036
EASTERN	0.238*	0.058	1.268	-0.082	0.055	0.922
MID_ATL	0.047	0.050	1.048	-0.229*	0.047	0.795
S_EAST	0.100	0.052	1.105	-0.270*	0.050	0.763
G_LAKES	0.036	0.054	1.036	-0.338*	0.052	0.713
S_WEST	0.183*	0.057	1.201	-0.208*	0.055	0.812
MID_CONT	0.105	0.064	1.111	-0.379*	0.062	0.685
ROCKIES	0.249*	0.066	1.283	-0.184*	0.064	0.832
WESTERN	0.110**	0.050	1.116	0.032	0.048	1.033
N	28,786.00			32,206.00		
% censored	41.1			43.1		
-2 log L	41209[a]			9585.4[a]		

Note: * = statistical significance at the 1 percent level; ** = statistical significance at the 5 percent level. See Table 3.1 for definitions of variables.

which the cohort is observed), the first specification for the FY92 cohort includes only months until the first and second promotions, while the second specification for the FY88 cohort includes months until the first, second, third, and fourth promotions. In the tables that follow, the variables representing months until each promotion are TPROM1, TPROM2, TPROM3, and TPROM4.

Two specifications are estimated because the inclusion of promotion speed might bias the results for the other variables. Insofar as there are unobservable characteristics that jointly determine promotion speed and retention, e.g., taste for public service, the estimated effects of the other covariates might be biased.

Whether higher-quality personnel stay longer in the DoD depends on the measure of quality used, on whether promotion speed is included in the model, and on the cohort. For the FY88 cohort, the results suggest that those who had better supervisor ratings and those who were promoted faster stayed longer. However, those who entered with more education did not always stay longer. The evidence in fact suggests that those who entered with the highest degrees had the poorest retention. On the other hand, the measurement problem associated with entry education makes it important to consider the other quality measures as well. The FY92 cohort results are less clear cut, making conclusions about the retention of higher-quality personnel difficult to reach.

FY88 Cohort Retention Results

When promotion speed is not included as a covariate in the model, the results indicate that those in the FY88 cohort with a better supervisor rating stayed longer in the DoD. That is, they had better retention. But when promotion speed is also included in the model, those with a better rating stayed for fewer months, i.e., they had poorer retention. More specifically, the results in the third and sixth column of Table 5.4 show that having another "outstanding" rating (a rating of 1) is estimated to reduce the hazard of separating from the DoD civil service—that is, increase retention—by 45.3 percent $(1 - 0.547)$ when promotion speed is not a covariate in the model, and by 50.6 percent when promotion speed is a covariate. The effect of having another "exceeds fully successful" rating (a rating of 2) is more modest. It is estimated to raise the separation hazard (reduce retention) by 41.5 percent when promotion speed is included but reduce it (increase retention) by 35.5 percent when promotion speed is not included. Thus, whether the results indicate that those in the FY88 cohort with better performance, as measured by supervisor rating, stayed longer depends on whether promotion speed is included in the model.

As noted in Chapters Two and Three, the estimated effects of the quality variables in the separation-hazard model reflect the better opportunities, both internal and external, available to higher-quality personnel. Thus, whether higher-quality personnel are retained depends on whether the retention effects of the internal opportunities exceed those of the external opportunities.

Promotion speed may affect the estimated impact of supervisor rating on retention because promotion speed is the outcome of the better opportunities available to higher-quality personnel inside the civil service (see Equations 2.3 and 2.4). Once the promotion-speed variables and therefore these better internal opportunities are incorporated into the model, the estimated effects of the other personnel quality measures, such as supervisor rating and education, reflect the better external opportunities available to higher-quality personnel. Thus, it is not surprising that, when account is taken of the outcome of their better internal civil service opportunities, those who perform better are estimated to leave sooner. When the internal opportunities are not incorporated in the model, i.e., when the promotion-speed variables are excluded, the estimates in Table 5.4 show that those who get better ratings stay longer. Also, not surprisingly, when internal opportunities are held constant (by incorporating the promotion-speed variables in the model), those with the highest rating and therefore the best external opportunities have a higher separation hazard than those whose rating is not quite as high. That is, the estimated separation hazard for individuals having another rating of 1 is 50.6 percent, while that for individuals having another rating of 2 is 41.5 percent. Thus, when internal opportunities are held constant, those with the best external opportunities are the most likely to leave.

Exclusion of promotion speed from the model provides some indication of whether the internal incentives are stronger than the external incentives, because internal incentives are not included in the set of control variables. The results for the FY88 cohort indicate that those who got better supervisor ratings had a stronger incentive to stay than to leave.

Inclusion of the promotion-speed variables also affects the estimated effects of the other quality measure–education at entry—in the analysis of retention. When promotion speed is included in the model, those with more education at entry have poorer retention. Specifically, having a bachelor's degree increases the separation hazard by 14.5 percent, while having a masters degree or a doctorate increases the hazard by 38.3 or 50 percent, respectively. Both estimated effects are statistically significant. Thus, when account is taken of the better internal opportunities available to higher-quality personnel, those with better external opportunities are found to have poorer retention.

However, even when promotion speed is excluded from the model, those in the FY88 cohort with more education were still sometimes found to have poorer retention, although the estimated effects are considerably smaller and not always statistically significant. The estimated effect of having a bachelor's degree on the FY88 separation hazard is –2.0, which is not statistically significant. Having a masters degree or a doctorate is estimated to increase the FY88 separation hazard (reduce retention) by 26.5 or 29.3 percent, respectively. Both of these estimated effects are statistically significant at the 1 percent level. They are also smaller than the 38.3 and 50.0 percent increases that are found when promotion speed is included in the model. Still, the fact that these estimated effects are positive means that even when no controls that capture the better internal opportunities available to better-educated personnel are included, these individuals stayed for shorter periods of time, other factors held constant. Thus, for the FY88 cohort, the internal opportunities did not provide a sufficiently strong incentive for the most-educated individuals to stay longer in the DoD civil service. These results are consistent with the promotion results, which indicated that those with the highest degrees were not always promoted faster than those with lower degrees.

The estimated effects of the final measures of quality, the promotion-speed variables, are relatively large and statistically significant.

Those who are promoted faster (i.e., those for whom the TPROM variables are numerically smaller) are estimated to have reduced separation hazards. That is, they have better retention. In the FY88 cohort, achieving the first promotion one month faster reduced the separation hazard by 5 percent. Achieving it 3 months faster reduced it by 15 percent. Achieving the second or third promotion one month faster reduced the separation hazard by about the same amount, 5.7 or 5.5 percent, respectively. Achieving the fourth promotion one month faster reduced the hazard by a much larger amount, 9 percent. Insofar as those who are promoted faster are better suited to the civil service and are of higher quality, these results suggest that when quality is measured by promotion speed, higher-quality personnel are retained longer.

Although the estimated effects of promotion speed are large, these estimates, like the other estimates shown in the first column of Table 5.4, may be biased because promotion speed and retention may be jointly determined. This might be the case, for example, if those with a stronger taste for the civil service perform better, get promoted faster, and are more likely to stay. It might also be the case if personnel managers strive to promote those individuals they think are the most likely to stay. Whatever the reason, if promotion speed and retention are jointly determined by some unobserved factor that is not included in the regression model, the estimated effects are biased, and the bias is likely to be negative. On the other hand, the magnitudes of the estimated effects are large, and even if biased, the true effects are likely to still be negative, although not as large.

Other factors were also found to affect the separation hazard for the FY88 cohort, although the estimated effects depend on whether the promotion-speed variables are included in the model. Most notably, those who entered at older ages, those who were supervisors or managers, and those in the Navy or Marine Corps were estimated to have a lower separation hazard, i.e., better retention. Those who entered in their 20s were estimated to have 30.1 percent lower hazard than the omitted group (those who entered above the age of 60), while those who entered in their 40s had a 54.4 percent lower separation hazard. The hazard for supervisors and managers was estimated to be about 2 percent lower than that for their nonsupervisor counterpart, while those in the Navy and Marine Corps were estimated to have separation hazards 11.8 and 14.6 percent lower, respectively. All of these estimated effects are statistically significant at the 1 percent level.

As shown in Figure 4.4, separation outcomes varied considerably by occupational area in the FY88 cohort. The results in Table 5.4 confirm that finding, even with other observable characteristics held constant. Those in science, mathematics, and engineering were estimated to have the lowest hazard rates; those in the medical field were estimated to have the highest hazard rates.

FY92 Cohort Retention Results

The results for the FY92 cohort are shown in Table 5.5. They differ somewhat from those for the FY88 cohort, although there are also similarities. As in the FY88 cohort, those in the FY92 cohort who were promoted faster stayed longer in the DoD civil service, and these effects are relatively large and statistically significant. Achieving the first promotion one month faster was estimated to reduce the separation hazard by 7.7 percent, while achieving the second promotion one month faster was estimated to reduce it by 14.7 percent. Again, however, the estimated effects of promotion speed may be biased downward, implying that the true effects may be smaller.

Achieving a second "outstanding rating" was estimated to increase the separation hazard (reduce retention) for the FY92 cohort by 110.7 percent when promotion speed is a covariate in the model and to increase it by only 6.7 percent when promotion speed is not a covariate. However, the latter estimated effect is not statistically significant at the 5 percent level. Achieving another "exceeds fully satisfactory" rating was estimated to increase the separation hazard by 105.1 percent when promotion speed is a covariate and increase it by only 20 percent when it is excluded. Both of these estimated effects are statistically significant at the 1 percent level. Thus, in contrast to the FY88 cohort, even when no account is taken of the outcome of the better internal opportunities available to those who perform better, these individuals in the FY92 cohort were found to have poorer retention.

Table 5.5

**Partial-Likelihood Cox-Regression-Model Estimates of Months
to Separation, FY92 Cohort**

	Includes Promotion Speed Variables			Excludes Promotion Speed Variables		
	Estimate	Std. Error	Risk Ratio	Estimate	Std. Error	Risk Ratio
CUMRAT1	0.745*	0.043	2.107	0.065	0.039	1.067
CUMRAT2	0.718*	0.045	2.051	0.183*	0.040	1.200
TPROM1	-0.080*	0.001	0.923			
TPROM2	-0.159*	0.003	0.853			
MNYOS0	-0.002*	0.001	0.998	-0.031*	0.001	0.970
CHAVRAT	-1.178*	0.057	0.308	-2.976*	0.055	0.051
DMDCVET	-0.453*	0.034	0.636	-0.036	0.033	0.965
SOMECOL0	0.009	0.029	1.009	-0.089*	0.027	0.915
AADEG0	-0.052	0.056	0.949	-0.131*	0.051	0.877
BADEG0	-0.043	0.037	0.958	-0.067**	0.034	0.935
ABOVBA0	0.143	0.092	1.154	-0.006	0.086	0.994
MA0	0.054	0.064	1.056	0.017	0.059	1.017
PHD0	0.112	0.159	1.118	-0.106	0.144	0.900
AGE20_0	0.320*	0.124	1.377	-0.022	0.111	0.978
AGE30_0	-0.040	0.117	0.961	-0.242**	0.105	0.785
AGE40_0	-0.305*	0.117	0.737	-0.401*	0.105	0.670
AGE50_0	-0.222	0.118	0.801	-0.637*	0.106	0.529
AGE60_0	-0.205	0.126	0.814	-0.407*	0.113	0.666
RACEMISS	0.660	1.003	1.935	-0.688	0.709	0.503
NONWHITE	-0.052**	0.025	0.950	-0.075*	0.023	0.927
FEMALE	0.024	0.028	1.024	0.121*	0.027	1.128
HCAPMIS0	-0.378*	0.135	0.686	-0.450*	0.129	0.638
HCAPCAT0	0.130**	0.056	1.138	0.023	0.051	1.024
EGRADE4	-0.188*	0.031	0.828	-0.258*	0.029	0.773
EGRADE5	-0.408*	0.038	0.665	-0.426*	0.035	0.653
EGRADE6	-0.630*	0.068	0.532	-0.582*	0.064	0.559
EGRADE7	-0.544*	0.055	0.580	-0.636*	0.051	0.529
EGRADE9	-0.365*	0.067	0.694	-0.542*	0.059	0.582
EGRADE11	-0.409*	0.075	0.664	-0.689*	0.069	0.502
EGRADE12	-0.345*	0.093	0.708	-0.461*	0.084	0.630
EGRADE13	-0.472*	0.146	0.624	-0.575*	0.142	0.563
EGRADE14	-0.623*	0.232	0.536	-0.757*	0.220	0.469
EGRADE15	-0.844	0.446	0.430			
FM10_0	-0.308**	0.129	0.735	-0.856*	0.120	0.425
FM11_0	-0.598*	0.096	0.550	-0.818*	0.086	0.441
FM20_0	-0.734**	0.313	0.480	-0.894*	0.297	0.409
FM21_0	0.089	0.099	1.093	0.202**	0.090	1.224
FM24_0	-0.256**	0.106	0.774	-0.231**	0.095	0.794

Table 5.5 (continued)

	Includes Promotion Speed Variables			Excludes Promotion Speed Variables		
	Estimate	Std. Error	Risk Ratio	Estimate	Std. Error	Risk Ratio
FM30_0	-0.359*	0.107	0.698	-0.584*	0.097	0.558
FM32_0	-0.501*	0.117	0.606	-0.546*	0.108	0.579
FM33_0	-0.339*	0.129	0.712	-0.529*	0.117	0.589
FM34_0	-0.181	0.096	0.834	-0.101	0.087	0.904
FM40_0	0.060	0.095	1.062	0.063	0.086	1.065
FM41_0	0.016	0.081	1.016	0.083	0.074	1.086
FM43_0	-0.264*	0.089	0.768	-0.295*	0.079	0.745
FM44_0	-0.108	0.070	0.898	-0.077	0.062	0.926
FM50_0	-0.218*	0.073	0.804	-0.028	0.065	0.972
FM51_0	-0.198**	0.084	0.820	-0.322*	0.076	0.725
FM52_0	-0.384*	0.085	0.681	-0.106	0.076	0.899
FM53_0	-0.264*	0.077	0.768	-0.186*	0.068	0.830
FM54_0	-0.141**	0.072	0.868	-0.130**	0.064	0.878
ARMY0	-0.080**	0.034	0.923	-0.186*	0.034	0.831
NAVY0	-0.201*	0.045	0.818	-0.091**	0.038	0.913
MARINE0	-0.085	0.114	0.919	-0.270*	0.084	0.763
AIRFORC0	-0.029	0.040	0.971	0.006	0.039	1.006
COMPET0	0.133*	0.028	1.143	0.215*	0.025	1.240
SUPMISS				1.226	0.713	3.407
SUP_MGR	-0.243*	0.064	0.784	-0.081	0.057	0.922
OPMMIS0	0.413*	0.070	1.511	0.117	0.060	1.124
NEWENG	0.209**	0.090	1.232	0.086	0.077	1.090
EASTERN	0.350*	0.083	1.419	-0.086	0.074	0.918
MID_ATL	0.230*	0.071	1.258	-0.238*	0.061	0.788
S_EAST	0.210*	0.073	1.234	-0.334*	0.063	0.716
G_LAKES	0.175**	0.075	1.191	-0.494*	0.068	0.610
S_WEST	0.176**	0.076	1.192	-0.186*	0.067	0.831
MID_CONT	0.315*	0.091	1.371	0.031	0.083	1.032
ROCKIES	0.245**	0.089	1.278	-0.235*	0.082	0.791
WESTERN	0.271**	0.072	1.311	0.046	0.062	1.047
N	17,398			19914		
% censored	49.88			49.7		
-2 log L	131499.48*			165666.64*		

Note: * = statistical significance at the 1 percent level; ** = statistical significance at the 5 percent level. See Table 3.1 for definitions of variables.

Also unlike the FY88 cohort, those in the FY92 cohort with more education were found to have a lower estimated separation hazard (better retention) when promotion speed is not included in the model. Those with a bachelor's degree were estimated to have a separation hazard 6.5 percent lower than those who had no college education. This estimated effect is statistically significant at the 1 percent level. The estimated effects of having education beyond a bachelor's degree are not statistically significantly different from zero. Thus, some evidence is found that those in the FY92 cohort with more education stayed longer, although the evidence is not overwhelming.

Other factors also affected the separation hazard for the FY92 cohort. Those with prior military service were estimated to have a 3.5 percent lower separation hazard than their nonveteran counterparts, and nonwhites were estimated to have a 7.3 percent lower separation hazard than their white counterparts. Those who were between 40 and 50 years of age at entry were estimated to have a separation hazard 33 percent lower than that for the omitted group, while entering in one's 20s was estimated to have a small, statistically insignificant effect on the separation hazard. As in the FY88 cohort, individuals in science, mathematics, or engineering were estimated to have the lowest separation hazards, other factors held constant. On the other hand, women were estimated to have a 12.8 percent higher separation hazard than men. Also as in the FY88 cohort, supervisors and managers had lower estimated hazard rates, as did those who worked for the Navy and the Marine Corps.

Chapter Six

CONCLUSIONS AND DIRECTIONS
FOR FUTURE RESEARCH

This report describes variations in promotion speed, retention, and pay among two recent cohorts of civil service personnel; develops proxy measures of personnel quality; and uses these measures to examine whether higher-quality GS workers are promoted faster, are retained in the DoD civil service longer, and are paid more than lower-quality workers are, and whether these patterns have changed in recent years.

To conduct this analysis, we constructed a longitudinal database that tracks through FY96 the careers of the workers who entered or reentered the DoD civil service between FY82 and FY96. The data capture information about entry characteristics and how they vary over each individual's career, as well as pay levels, promotion events and timing, and the timing of exits from the DoD. By tracking the careers of two cohorts, the database permits a comparison across occupational groups and time periods, including a period covering the DoD drawdown. The analysis focuses on the FY88 cohort, whose members entered or reentered prior to the defense downsizing, and on the FY92 cohort, whose members entered or reentered during the defense downsizing.

The data also permit the construction of proxy measures of personnel quality. Three measures are used: entry education, supervisor ratings, and promotion speed. Promotion speed is studied both as an indicator of personnel quality and as a personnel outcome. It is used as a personnel quality indicator in the analysis of whether higher-quality personnel are retained. Because each of the three measures has its advantages and its drawbacks, more than one measure is used. A regression framework is used to control for other factors that may affect promotion, pay, and retention.

MAIN CONCLUSIONS

The analysis of pay for both cohorts suggests that higher-quality personnel are generally paid more. Those who receive the highest supervisor ratings and thus presumably perform better in their jobs are found to be paid more than those who receive the lowest ratings, when other observed characteristics are held constant. The increment in annual pay associated with receiving the top rating ranges from 2.8 to 4.3 percent. The analysis also suggests that those with more education earn more in the DoD civil service. Furthermore, the relationship between pay and education may be underestimated because the pay variable excludes bonuses and special pays, an omission that could produce a downward bias.

The analysis also indicates that higher-quality GS personnel are generally promoted faster. Those who get more top ratings in their supervisor appraisals are promoted in fewer months than those who get lower ratings. Furthermore, the higher the supervisor rating, the faster the estimated promotion speed. Those with any college education are also promoted faster than those with no college education, although having more education than a bachelor's degree does not always translate into faster promotion.

The analysis of retention in the DoD civil service produces less clear-cut results, especially for the FY92 cohort. The results on retention of higher-quality personnel depend on the measure of quality used, on whether promotion speed is included in the model, and on the cohort. For the FY88 cohort, the analysis indicates that those who had better supervisor ratings and who were promoted faster stayed longer in the DoD. That is, those who were better matched to the civil service, as captured by their supervisor rating and promotion speed, had a stronger incentive to stay than to leave. However, when personnel quality is measured in terms of entry education, those in the FY88 cohort with at least a bachelor's degree were found to have had poorer retention, although the estimated effects are not always statistically significant and may be biased due to measurement error. Thus, for the FY88 cohort, the internal opportunities for those with advanced degrees do not appear to have provided a sufficiently strong incentive to stay longer in the DoD civil service.

The results for entry education and supervisor rating differ somewhat for the two cohorts. Some evidence is found that, unlike the members of the FY88 cohort, those with any college in the FY92 cohort stayed longer in the DoD, although the evidence is not overwhelming because the results are statistically significant only for those with a bachelor's degree or less, not for those with more education. Thus, for the FY92 cohort, internal incentives to stay in service appear to have been stronger than the incentives to leave for those with an associate's or bachelor's degree, but not necessarily for those with a masters degree or a doctorate. The evidence also does not suggest that those in the FY92 cohort who got better supervisor ratings stayed longer, although there is some indication that some of the better performers, particularly those who received a rating of 2 (the second-highest score), had poorer retention than those who had the lowest ratings.

The estimated retention effects of the final measure of quality, promotion speed, are relatively large and statistically significant for the FY92 cohort. Those who were promoted faster are estimated to have had significantly better retention. However, the magnitude of the estimated effects is likely to be biased for both the FY92 and FY88 cohorts because promotion speed and retention may be jointly determined, and the mechanism that jointly determines them is not incorporated in the model.

The analysis of retention in the DoD civil service provides some evidence, although it is not overwhelming, that higher-quality personnel stay longer, especially when quality is measured in terms of having faster promotions and, to some extent, in terms of having better supervisor ratings. This is especially true for the FY88 cohort, which represents the status quo prior to the commencement of the huge DoD downsizing that occurred in the 1990s. But the analysis also suggests that those with the most education—masters degrees or doctorates—do not necessarily seem to stay longer in the DoD.

The regression results for the two cohorts also give some indication of the degree to which promotion, pay, and retention vary among DoD civil service personnel. The analysis, although primarily descriptive, indicates that these profiles vary considerably across occupational areas, even after other observed individual and job characteristics are accounted for. Entry pay and pay growth differ considerably across occupational areas, as do promotion speed and retention. These results suggest that careers vary significantly across occupational areas despite the common pay table that serves all GS personnel in all occupations. They also suggest that personnel managers may have more flexibility in varying the pay of personnel in different occupations than the common pay table would lead one to expect, because promotion speed can vary considerably.

On the other hand, just because promotion speed can operate to provide pay raises, it is not clear this is the best or most efficient means of raising pay. If personnel are well-suited for their current job but are unsuited for higher-ranking positions, promoting them in order to give them a sizable pay raise may not be an efficient policy. It might be more efficient to enable personnel managers to give sizable pay raises without promotions, abstracting from current personnel policy such as the classification system. Whether such a policy would in fact be more efficient depends on the extent to which performance across ranks is positively or negatively correlated. Those who perform well in the lower ranks of the civil service may also be the best performers in the upper ranks. Thus, promotion as a means of providing pay raises may be a sensible policy. The efficiency of the policy also depends on the incentives of the managers to identify and give sizable raises only to those with the best performance. Since civil service managers are not profit maximizers or residual claimants the way private-sector owners are, they may even elect to give everyone, regardless of quality, large pay raises as a means of maintaining a happy workforce.

Identifying the most efficient set of compensation and personnel policies to ensure that better performers and higher-quality personnel are paid more is beyond the scope of this analysis. The analysis does suggest that current policies generally operate to pay and promote better GS workers, but these policies may not be sufficient to retain them, especially those with the most education. Another analysis that uses the same data (Gibbs, forthcoming) indicates that better-educated scientists and engineers in the DoD laboratories have higher retention. Thus, the poorer retention of those with more education, particularly those with advanced degrees, must occur in nonlaboratory settings and in other occupations.

AREAS FOR FUTURE RESEARCH

Although this analysis provides some answers, other questions remain. Therefore, future research might consider addressing the following questions:

- To what extent does promotion speed identify better performers? Promotion speed is determined in part by vacancy rates or, more generally, by requirements for personnel in the upper ranks. This analysis does not address the extent to which promotion speed reflects differential vacancy rates, although account is taken of occupation, region, and other factors that might determine those rates. Because vacancy rates can vary within occupations and among regions, future research should attempt to take better account of the role of vacancy rates in determining the speed of promotion of higher-quality personnel.

- Is the retention of higher-quality personnel sufficient to meet current and future personnel requirements? Although some (not overwhelming) evidence was found to indicate that higher-quality personnel stay longer, whether this retention is sufficient to meet existing and future requirements is still an open question. Future research should examine whether the supply of higher-quality personnel is sufficient.

- Are better measures of personnel quality available? Such measures might include test scores and explicit measures of productivity. Future research should attempt to refine the measures of personnel quality.

- To what extent do higher-quality personnel defer retirement? This analysis focused on retention of new and midcareer personnel. As the civil service workforce ages, along with the rest of the baby boom generation, a greater fraction of that workforce will be near retirement age. Analysis should be conducted on whether higher-quality personnel choose to retire later or earlier than lower-quality personnel. That is, future analysis should address the separation decisions of older personnel.

- Do bonuses and special pays provide effective inducements for higher-quality personnel to stay in the DoD civil service? Available evidence indicates that bonuses are effective in improving retention in specific occupations and geographical areas (U.S. Office of Personnel Management, 1999). However, it is not known whether they are differentially more effective for higher-quality than for lower-quality personnel.

More generally, greater use should be made of available data on civil service personnel and on the dataset constructed for this analysis. As more is understood about civil service careers, more equitable and more cost-effective compensation and personnel policies can be put in place to manage the federal civil service workforce.

OCCUPATIONS INCLUDED IN THE ANALYSIS

Table A.1 shows the occupations represented in each occupational group in the analysis presented in this report. The table lists the occupational groups (including their MOG-FOG coding) and the specific occupations included in them. The four-digit occupational codes are also shown.

Table A.1

Major Occupations in Each MOG-FOG Occupational Group

10 : Science
0180 Psychology
0401 General biological science
1301 General physical science
1310 Physics
1320 Chemistry
1350 Geology
1370 Cartography

11: Engineering
0801 General engineering
0808 Architecture
0810 Civil engineering
0819 Environmental engineering
0830 Mechanical engineering
0840 Nuclear engineering
0850 Electrical engineering
0855 Electronics engineering
0861 Aerospace engineering
0893 Chemical engineering
0896 Industrial engineering

20: Mathematics
1515 Operations research
1520 Mathematics
1529 Mathematical statistician
1530 Statistician

21: Medical
0601 General facilities and equipment
0602 Medical officer
0610 Nurse
0630 Dietitian and nutritionist
0644 Medical technologist
0660 Pharmacist
0690 Industrial hygiene

22: Legal
0904 Law clerk
0905 General attorney
0950 Paralegal specialist
1222 Patent attorney

23: Education
1701 General education and training
1710 Education and vocational training
1740 Education services

24: Miscellaneous professional
0028 Environmental protection specialist
0101 Social science
0132 Intelligence
0185 Social work
1035 Public affairs
1410 Librarian
1550 Computer science

30: Logistics Management
0018 Safety and occupational health management
0346 Logistics management
1101 General business and industry
1102 Contracting
1150 Industrial specialist
2001 General supply
2003 Supply program management
2010 Inventory management
2101 Transportation specialist

31: Personnel Management
0201 Personnel management
0205 Military personnel management
0212 Personnel staffing
0221 Position-classification
0230 Employee relations
0235 Employee development
0260 Equal employment opportunity

32: Financial Management
0501 Financial administration and program
0510 Accounting
0511 Auditing
0560 Budget analysis

33: Data Systems Management
0033

34: Central Management
0127
0032 Intelligence
0203 Personnel clerical and assistance
0029 Environmental protection assistant
0035
0086
0033
0021 Community planning technician
0049

40: Science and Engineering Technician
0802 Engineering technician
0817 Surveying technician
0818 Engineering drafting
0856 Electronics technician
1311 Physical science technician

41: Medical Technician
0620 Practical nurse
0640 Health aide and technician
0645 Medical technician
0647 Diagnostic radiologic technologist
0675 Medical records technician
0681 Dental assistant

42: Logistics Technician
1152 Production control
1670 Equipment specialist
1910 Quality assurance
2005 Supply clerical and technician
2099 Supply student trainee
2154 Air traffic assistant

43: Management Technician
0332 Computer operation
0335 Computer clerk and assistant
0344 Management and program clerical and assistance
0525 Accounting technician

44: Miscellaneous Technician
0099 General student trainee
0189 Recreation aide and assistant
0899 Engineering and architecture student trainee
1702 Education and training technician

50: Secretary
0318 Secretary
0322 Clerk-typist

51: Financial Clerk
0503 Financial clerical and assistance
0530 Cash processing
0540 Voucher examining
0544 Civilian pay
0545 Military pay
0561 Budget clerical and assistance

52: Logistics Clerk
1105 Purchasing
1106 Procurement clerical and technician
2005 Supply clerical and technician
2134

53: General Office Operations
0203 Personnel clerical and assistance
0204 Military personnel clerical and technician
0305 Mail and file
0356 Data transcriber

54: Miscellaneous Clerical
0303 Miscellaneous clerk and assistance
0399 Administration and office support student trainee
0679 Medical clerk
1411 Library technician
2102 Transportation clerk and assistant

60: Medical Attendants
0621 Nursing assistant
0661 Pharmacy technician
0699 Medical and health student trainee

61: Fire/Police
0081 Fire protection and prevention
0083 Police
0085 Security guard

62: Personnel Services
2091 Sales store clerical

Advisory Committee on Federal Workforce Quality Assessment, "Federal Workforce Quality: Measurement and Improvement," A Report to the Washington, D.C.: U.S. Merit Board and the U.S. Office of Personnel Management, U.S. Merit Systems Protection Board, 1992.

Allison, Paul D., *Survival Analysis Using the SAS System: A Practical Guide*, Cary, NC: SAS Institute Inc., 1995.

Asch, Beth, and John Warner, Separation and Retirement Incentives in the Civil Service: A Comparison of the Federal Employees Retirement System and the Civil Service Retirement System., Santa Monica, CA: RAND, MR-986-OSD, 1999.

_____, *A Theory of Military Compensation and Personnel Policy*, Santa Monica, CA: RAND, MR-439-OSD, 1994.

Buddin, Richard, Daniel S. Levy, Janet Hanley, and Donald M. Waldman, *Promotion Tempo and Enlisted Retention*, Santa Monica, CA: RAND, R-4135-FMP, 1992.

Committee on Scientists and Engineers in the Federal Civil Service, *Improving the Recruitment, Retention, and Utilization of Federal Scientists and Engineers*, Alan K. Campbell, Stephen J. Lukasik, and Michael G. McGeary (eds.), National Research Council, Washington, D.C.: National Academy Press, 1993.

Gibbs, Michael, "Pay Competitiveness and Quality of Civil Service Scientists and Engineers in DoD Laboratories," Santa Monica, CA: RAND, forthcoming.

Hartman, Robert, *Pay and Pensions for Federal Workers*, Washington, D.C.: The Brookings Institution, 1983.

Johnston, William, *Civil Service 2000*, The Hudson Institute, Prepared for the U.S. Office of Personnel Management, Career Entry Group, Washington, D.C., 1988.

Kettl, Donald F., Patricia W. Ingraham, Ronald P. Sanders, and Constance Horner, *Civil Service Reform: Building a Government that Works*, Washington, D.C.: The Brookings Institution, 1996.

Mace, Don, and Erik Yoder, "Federal Employees Almanac, 1998," *Federal Employees News Digest*, 1999.

Orvis, Bruce, James Hosek, and Michael Mattock, *PACER SHARE Productivity and Personnel Management Demonstration: Third Year Evaluation*, Santa Monica, CA: RAND, MR-310-OSD, 1993.

U.S. Office of Personnel Management, Office of Merit Systems Oversight and Effectiveness, *Report of a Special Study, The 3Rs: Lessons Learned from Recruitment, Relocation, and Retention Incentives*, Washington, D.C., December 1999.